T0406600

Wolfgang Seeger

Endoscopic and Microsurgical Anatomy of the Cranial Base

In collaboration with
Jan Kaminsky and Astrid Weyerbrock

SpringerWienNewYork

Prof. em. Dr. med. Wolfgang Seeger
Department of Neurosurgery, University Hospital Freiburg, Freiburg/Br.,
Germany

© 2010 Springer-Verlag/Wien
Printed in Austria

SpringerWienNewYork is a part of Springer Science+Business Media
springer.at

Typesetting and Printing:
Druckerei Theiss GmbH, St. Stefan, Austria, www.theiss.at

Printed on acid-free and chlorine-free bleached paper

SPIN: 12641352

With 79 (partly coloured) Figures

Library of Congress Control Number: 2009936964

ISBN 978-3-211-99319-4 SpringerWienNewYork

Preface

Neuroendoscopic transnasal surgical approaches have become increasingly common for some time, and have started to replace microsurgical techniques in pituitary surgery. Endoscopic transnasal approaches have also been recently used in some neurosurgical centres to reach pre- and retrosellar areas and targets localized in the basal cisterns. One major factor is the localization of the carotid artery between the siphon and aperture externa of the carotid canal at the base of the petrous bone.

The following areas are especially critical:

1. the anterior siphon area
2. the area between the bottom of the sphenoid sinus and the base of Processus pterygoideus
3. the proximity of the carotid canal to the Tuba Eustachii, the labyrinth, Bulbus superior of the internal jugular vein and the facial nerve.

Some segments of the course of the carotid artery are well known but they have rarely been surgical target areas using transnasal approaches to date.

This has changed since the introduction of modern imaging techniques, especially neuronavigation. It has become possible to identify and localize structures of the skull and extra- and intracranial structures in any desired plane.

The study of the anatomy of the skull base should not exclude experience gained in cadaver skull dissections. In contrast to imaging techniques, standard cadaver skull dissections do not permit examination of the above mentioned critical areas of the skull base without destruction of the skull specimen. Therefore it is necessary to also present a representative variety of skull preparations, which this book seeks to do.

The most important blood vessels and nerves passing the bony foramina, foveae and fissures have been labelled by colour. As this book is based on a collection of skulls, rare or unknown anatomical variants may be illustrated which are commonly not or only rarely found in anatomical atlases or standard medical textbooks. One example of a rare anatomic variant is the variability of the foramen spinosum and the adjacent Spina angularis, which are covered by the Tuba auditiva and the origins of M. tensor and the levator veli palatini. These variants might gain some relevance for neuronavigation and endoscopy because of the closeness of A. meningea media and the transitional area between the Pars ossea and Pars cartilaginea tubae.

The author owes a particular debt of gratitude to the chairs of the departments of neurosurgery in Freiburg (Prof. Zentner) and Giessen (Prof. Böker), and to the director of the neurosurgical department in Zwickau (Prof. Warnke), and their coworkers, especially PD Dr. Kaminsky (Freiburg), Dr. Nestler, Dr. Preuss, and the oto-rhino-laryngologist PD Dr. Bockmühl (Giessen).

Dr. Astrid Weyerbrock has carefully revised and edited the manuscript.

The author is grateful to Ms. E. Rotermund, Professor Zentner's secretary, for the typing and layout of the text.

I would especially like to thank the Springer Verlag Wien New York for its continuous support and cooperation and excellent publication of my books for over 3 decades.

Freiburg i. Br., October 2009   *Wolfgang Seeger*

Contents

CHAPTER I
SURVEY
(Figs. 1 to 8)

Overview (Figs. 1 to 5)

The following figures illustrate the anatomical base of transnasal endoscopic routes but not according to conventional anatomical views.
For better understanding, it is useful to give an overview of the well known anatomy as shown in Figs. 1 to 5.

Topography of the endoscopic routes (Figs. 6 to 8)

The routes can be divided into five segments:
Cavum nasi and paranasal sinuses
Foramen sphenopalatinum and adjacent structures
Foramen lacerum and adjacent structures
Pyramis (petrous bone)
Area of Clivus

Cavum nasi and paranasal sinuses

Unusual and nearly unknown routes of CSF leaks are important, which occur more frequently in extended transnasal endoscopy (Castelnuovo P, Locatelli D, 2007).
Other aspects of the transnasal routes are well known from pituitary surgery.

Foramen sphenopalatinum and adjacent structures

Foramen sphenopalatinum is located at the bony connection of the anterior inferior segment of Sinus sphenoidalis, Meatus nasi medius, and Fossa pterygopalatina. Its dorsal margin corresponds to the anterior Apertura of Canalis pterygoideus.

Foramen lacerum

Its anterior margin corresponds to the posterior Apertura of Canalis pterygoideus. The Canalis penetrates the base of Processus pterygoideus (for definition of the course of the carotid artery between Sinus sphenoidalis and Foramen lacerum). Canalis pterygoideus is an important landmark at surgery, as it defines the course of the carotid artery …

Pyramis (petrous bone)

Anterior area: A. carotis int. and its relationship to Tuba and A. meningea media
Posterior area: A. carotis int. and its relationship to Labyrinth, Tuba, Foramen jugulare and Canalis n. hypoglossi.

SURVEY (Figs. 1 to 8)

Fig. 1

Overview

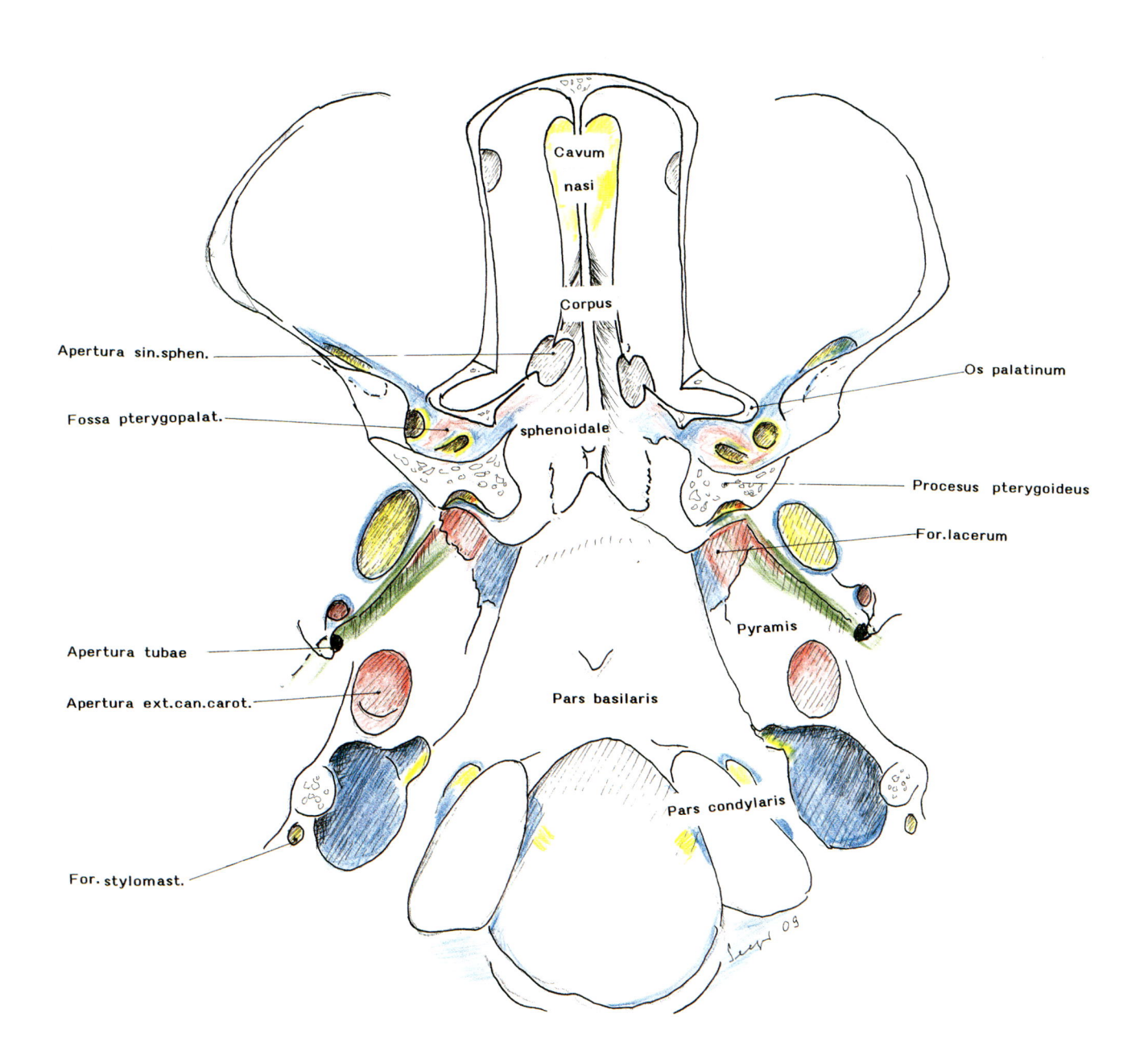

Cavum
nasi

Corpus

Apertura sin.sphen.

Os palatinum

Fossa pterygopalat.

sphenoidale

Procesus pterygoideus

For.lacerum

Apertura tubae

Pyramis

Apertura ext.can.carot.

Pars basilaris

Pars condylaris

For. stylomast.

Fig. 2

Cranial base. Extracranial medial area

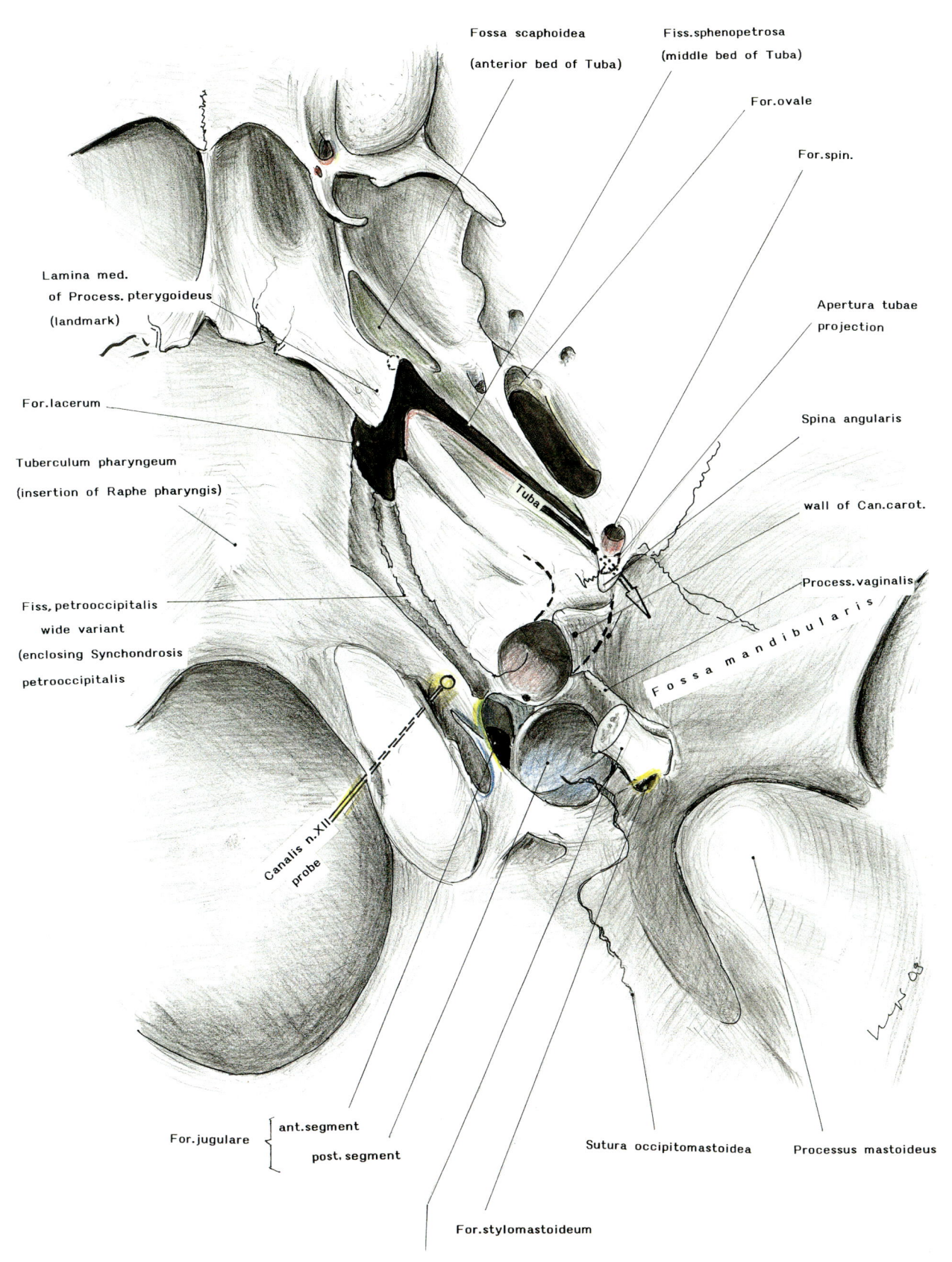

Fossa scaphoidea
(anterior bed of Tuba)

Fiss.sphenopetrosa
(middle bed of Tuba)

For.ovale

For.spin.

Lamina med.
of Process. pterygoideus
(landmark)

Apertura tubae
projection

Spina angularis

For.lacerum

Tuba

wall of Can.carot.

Tuberculum pharyngeum
(insertion of Raphe pharyngis)

Process.vaginalis

Fossa mandibularis

Fiss, petrooccipitalis
wide variant
(enclosing Synchondrosis
petrooccipitalis

Canalis n.XII
probe

For.jugulare { ant.segment
                post. segment

Sutura occipitomastoidea

Processus mastoideus

For.stylomastoideum

Process.styl.

Fig. 3

Addendum to Fig. 2
Cranial nerves and blood vessels, schematic

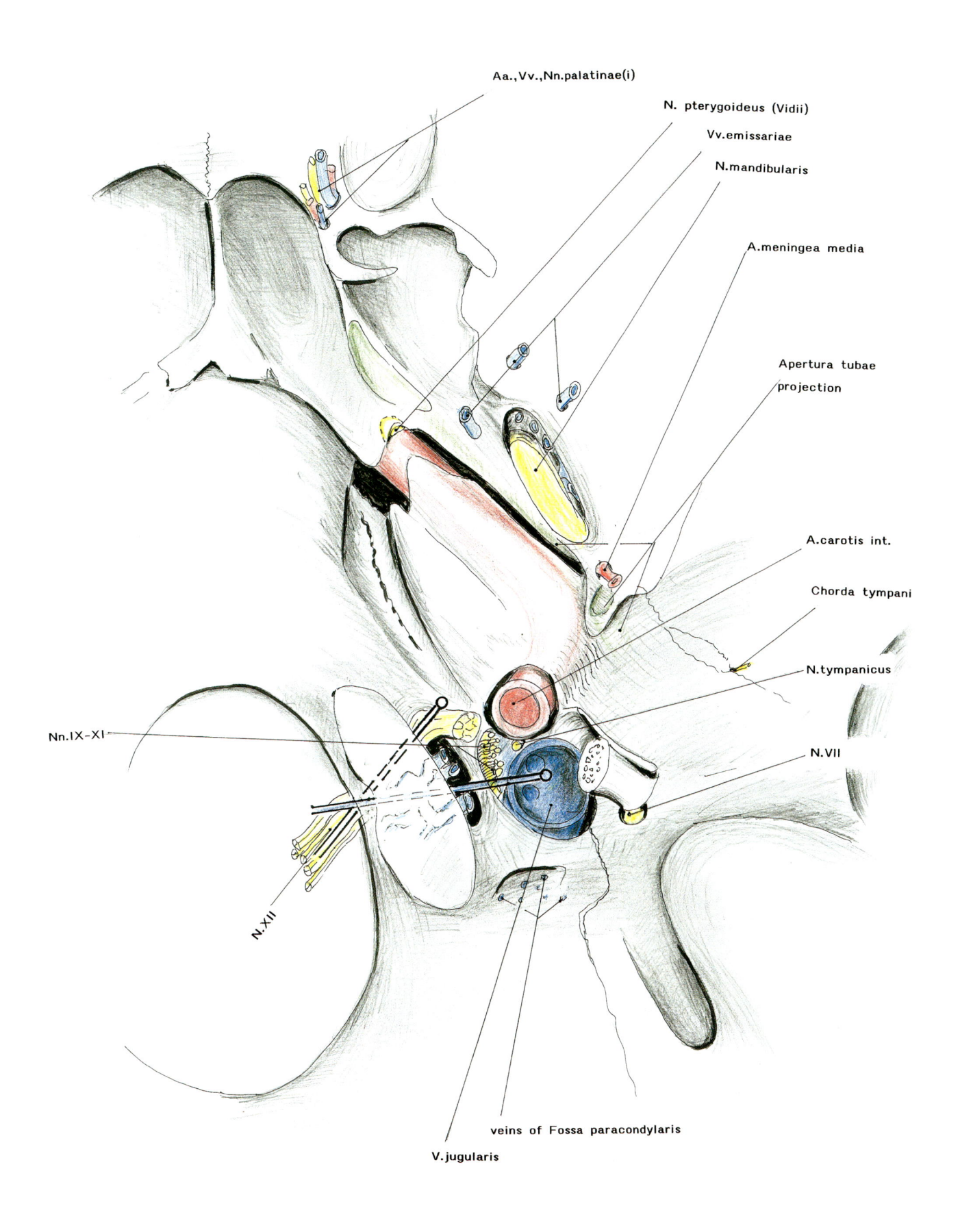

Aa.,Vv.,Nn.palatinae(i)

N. pterygoideus (Vidii)

Vv.emissariae

N.mandibularis

A.meningea media

Apertura tubae projection

A.carotis int.

Chorda tympani

N.tympanicus

N.VII

Nn.IX–XI

N.XII

veins of Fossa paracondylaris

V.jugularis

Fig. 4

Skull base. Intracranial medial area.

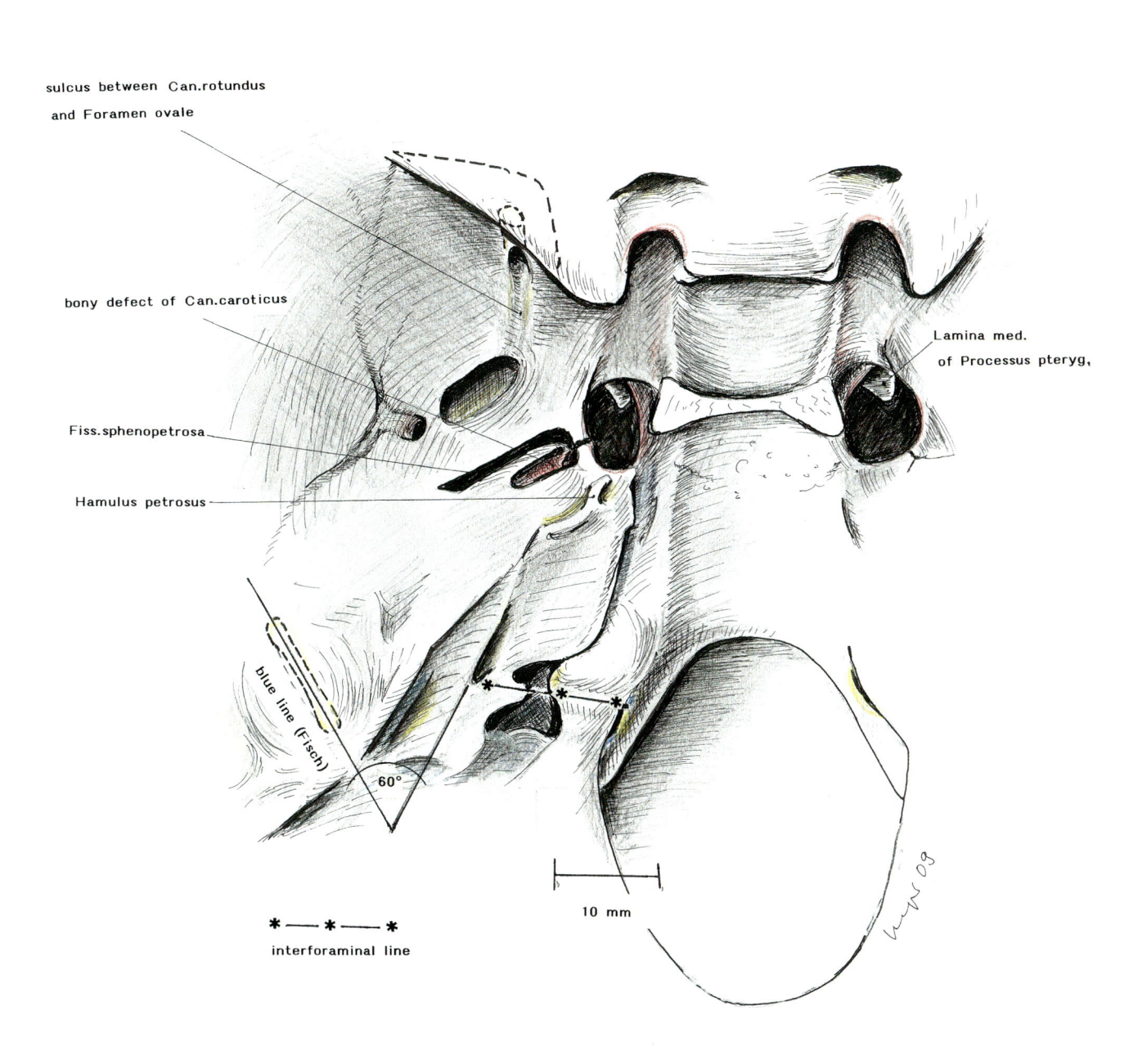

sulcus between Can.rotundus
and Foramen ovale

bony defect of Can.caroticus

Fiss.sphenopetrosa

Hamulus petrosus

blue line (Fisch)

60°

Lamina med.
of Processus pteryg,

10 mm

* — * — *
interforaminal line

Fig. 5

Addendum to Fig. 4
Cranial nerves and blood vessels, schematic

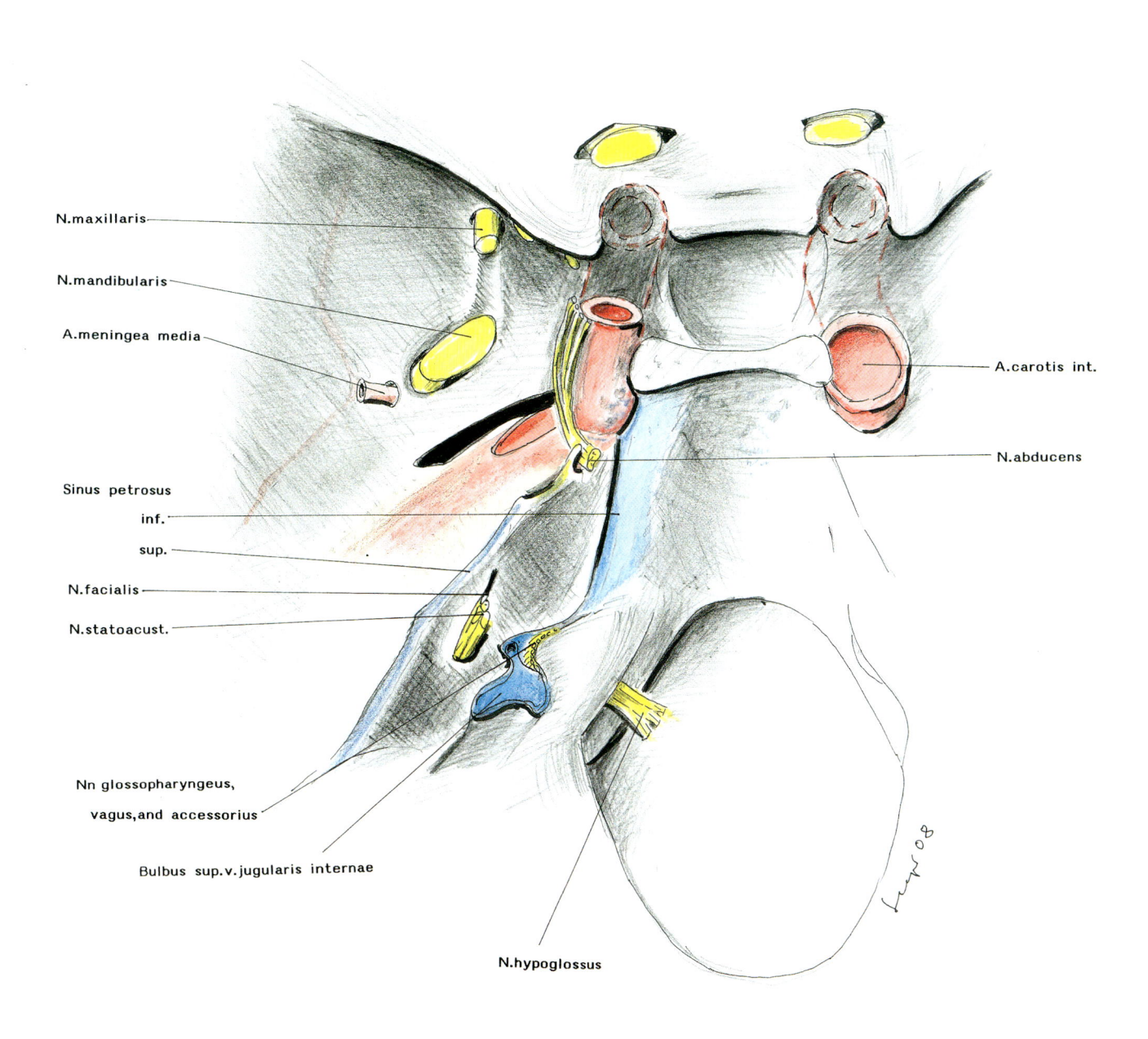

N.maxillaris

N.mandibularis

A.meningea media

A.carotis int.

N.abducens

Sinus petrosus

    inf.

    sup.

N.facialis

N.statoacust.

Nn glossopharyngeus,
vagus, and accessorius

Bulbus sup.v.jugularis internae

N.hypoglossus

Fig. 6

Cadaver skull dissection for better understanding of transnasal endoscopic approaches

The upper areas of Cavum nasi and Orbita are transected. The base of Processus ptery-goideus is transected to illustrate the relationship of Cavum nasi (dorsal area of Choana) to the roof of Foramen sphenopalatinum and Fossa pterygopalatina. These are located anterior to Processus pterygoideus.
Foramen lacerum is located posterior to Processus pterygoideus. Fossa pterygopalatina and the area of Foramen lacerum (+ Apertura int. of Canalis caroticus) are connected by Canalis pterygoideus (Vidii) (projection, dotted).
The middle and inferior segment of Pyramis (Pars petrosa of the temporal bone) con-tain structures which are located close to the Labyrinth.

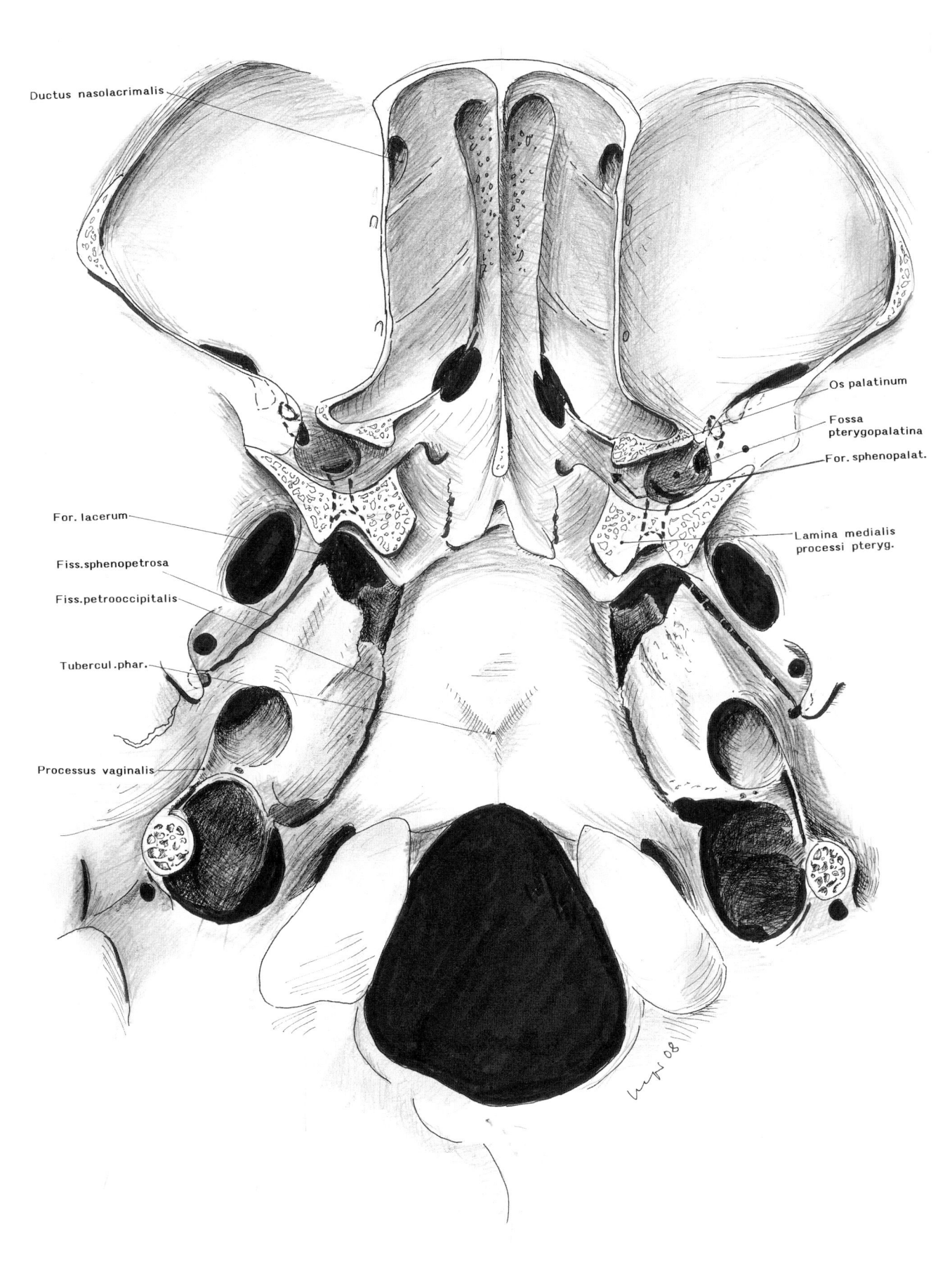

Ductus nasolacrimalis

Os palatinum

Fossa pterygopalatina

For. sphenopalat.

For. lacerum

Fiss.sphenopetrosa

Fiss.petrooccipitalis

Lamina medialis processi pteryg.

Tubercul.phar.

Processus vaginalis

Fig. 7

Roof at Foramen sphenopalatinum and surrounding structures.
Sectional magnification of Fig. 3, slightly modified.

Foramen sphenopalatinum is located between Fossa pterygopalatina and Cavum nasi.
Its posterior limit area is Os palatinum. It is partially or completely fused with Processus pterygoideus. The base of Processus pterygoideus is interposed between Foramen sphenopalatinum and Foramen lacerum.
The Processus-pterygoideus-Corpus-sphenoidale-bloc (merged perinatally) contains Canalis pterygoideus (Vidii) (dotted). It connects the anterior margin of Foramen lacerum (area of the carotid artery) to Fossa pterygopalatina.

Canalis pterygoideus is an important landmark for the endoscopic visualization of A. carotis interna from the lateral wall of Sinus sphenoidalis to Foramen lacerum, Apertura interna of Canalis caroticus, and to all segments of Canalis caroticus.*

---

* according to Kassam, Snyderman et al (2005), and according to the surgical experiences of Kaminsky, Freiburg/Br., personal communication 2009.

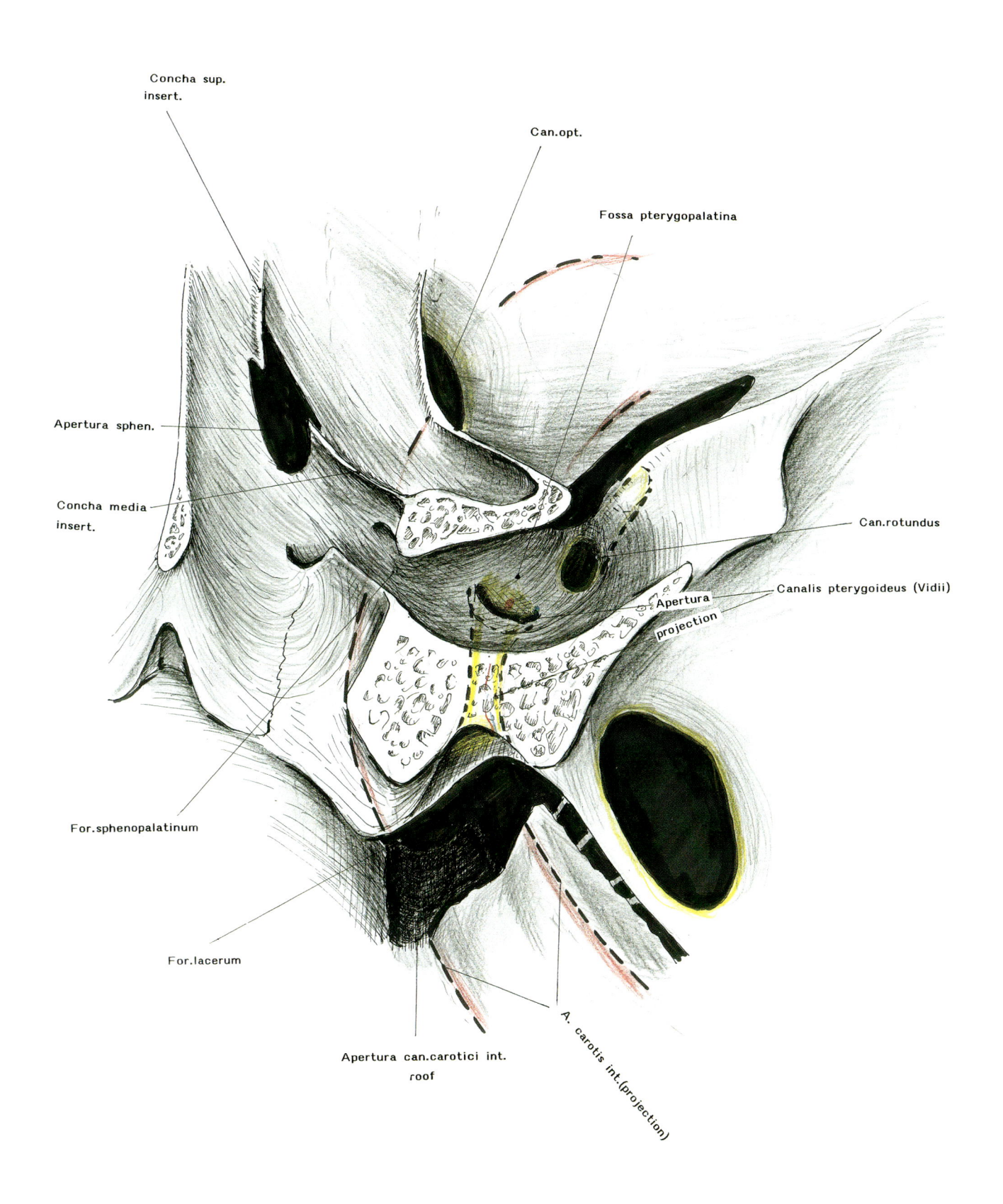

Concha sup. insert.

Can.opt.

Fossa pterygopalatina

Apertura sphen.

Concha media insert.

Can.rotundus

Apertura projection

Canalis pterygoideus (Vidii)

For.sphenopalatinum

For.lacerum

Apertura can.carotici int. roof

A. carotis int.(projection)

Fig. 8

Addendum to Fig. 7
Course of A. carotis int. between Apertura interna and ext. of Canalis caroticus.
Variant of Apertura ext.

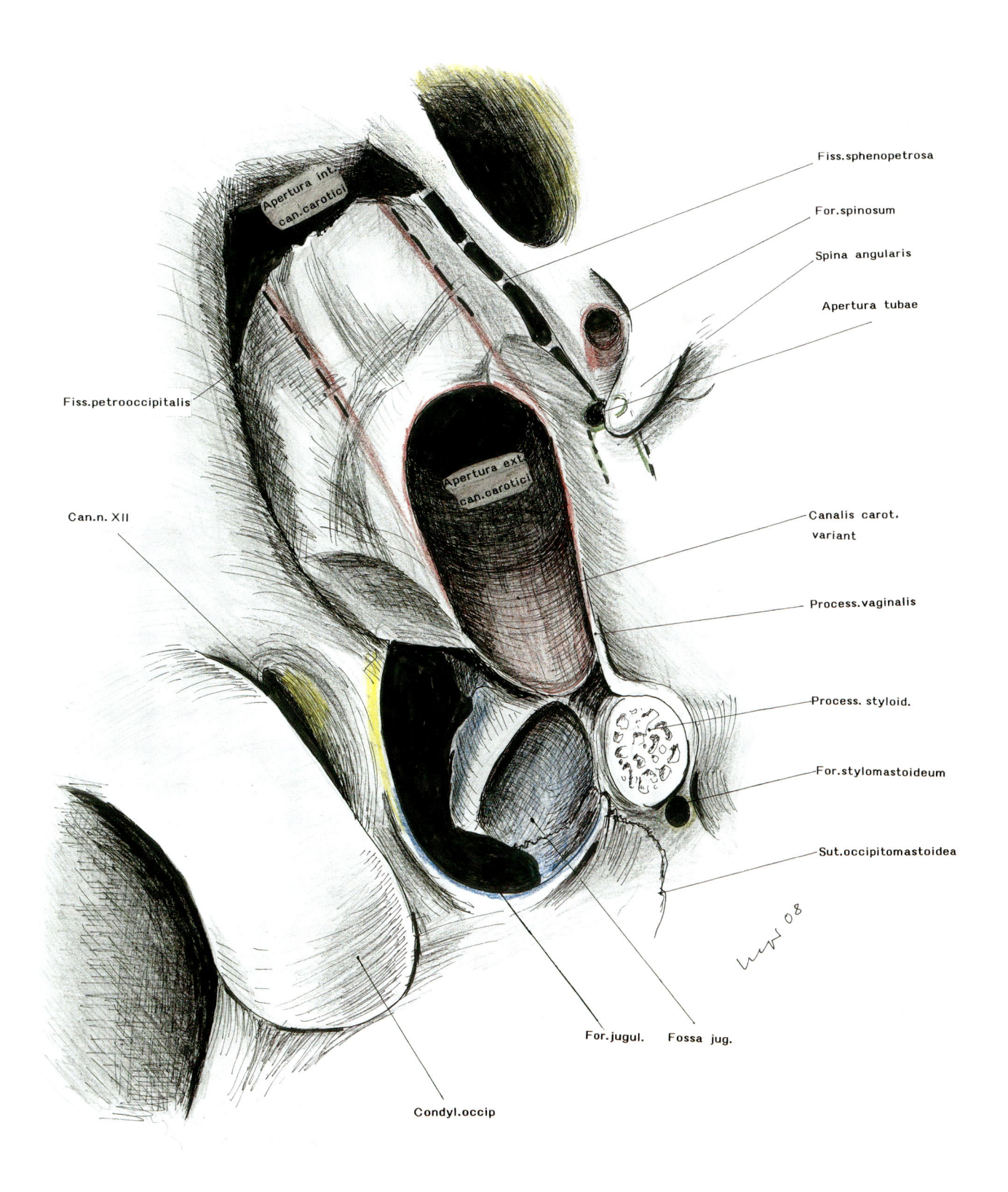

Fiss.sphenopetrosa

For.spinosum

Spina angularis

Apertura tubae

Apertura int.
can.carotici

Fiss.petrooccipitalis

Apertura ext.
can.carotici

Canalis carot.
variant

Can.n. XII

Process.vaginalis

Process. styloid.

For.stylomastoideum

Sut.occipitomastoidea

For.jugul.   Fossa jug.

Condyl.occip

Literature

Feneis II (1982) Anatomisches Bildwörterbuch der internationalen Nomenklatur, 5. Aufl. Thieme, Stuttgart

Kassam AB, Gardner P, Snyderman C, Mintz A, Carrau R (2005) Expanded endonasal approach: The rostrocaudal axis. Part II. Posterior clinoides to the foramen magnum. Neurosurg Focus Jul 15: 19(1): E4

Lang J (1979) Kopf, Teil B, Gehirn- und Augenschädel. Springer, Berlin Heidelberg New York

Rauber-Kopsch (1906) Lehrbuch der Anatomie, Bd. 2. Georg Thieme, Leipzig

CHAPTER II
CAVUM NASI AND
FOSSA PTERYGOPALATINA
(Figs. 9 to 17)

Overview (Figs. 9 and 10)

The superior transnasal routes for endoscopy are Cavum nasi, Sinus sphenoidalis, the roof of Os ethmoidale, and Sella turcica.

The inferior transnasal routes for endoscopy are Cavum nasi, Sinus sphenoidalis, Cavum nasi inferior and Fossa pterygopalatina, along A. carotis interna.

Cavum nasi (Fig. 10)

The route crossing Meatus nasi medius is well known in pituitary surgery. Connections to paranasal sinuses and Ductus nasolacrimalis are responsible for infections and CSF leaks.

Hiatus maxillaris (Fig. 11)

It is usually located at Infundibulum ethmoidale, anterior to Bulla ethmoidalis. Bulla ethmoidalis is the largest ethmoid cell. Additional fenestrations may be present as bony gaps around Bulla ethmoidalis. If Mucosa is fenestrated at these points, additional Hiati maxillaris may be visible. These may present further entry zones for infections of Sinus maxillaris.

Other paranasal sinuses (Fig. 12)

All paranasal sinuses are draining to Meatus nasi superior or medius. Meatus nasi inferior is separate and only drains Ductus nasolacrimalis. Besides the well known routes of CSF leaks via the paranasal sinuses, other routes must taken in consideration, which were rare in the past, but are more common today.

Atypical CSF leaks through the roof of Orbita
Usually CSF leaks at the orbital roof are draining into Sinus frontalis. Atypical extensions of the anterior or posterior ethmoid cells into the orbital roof may present atypical routes of CSF leaks.

Pneumatization of the anterior clinoid process is well known.
CSF leaks may occur after conventional and endoscopic surgery. Its route may be a connection to posterior ethmoid cells or to Sinus sphenoidalis between Canalis opticus and Planum sphenoidale (duplication of the roof of Canalis opticus), as described by Lang (1981) and by Seeger (2003).

CSF leaks from Canalis rotundus
to Sinus sphenoidalis are possible, because the wall of Canalis rotundus and the Sulcus between Foramen ovale and Canalis rotundus are segments of the wall of Sinus sphenoidalis (Seeger 2000)

Foramen sphenopalatinum and Fossa pterygopalatina (Figs. 13 to 19)

Foramen sphenopalatinum

This foramen is interposed between the posterior segment of Meatus nasi medius, the inferior segment of Sinus sphenoidalis, and the medial segment of Fossa pterygopalatina (Fig. 14)

Fossa pterygopalatina

This fossa extends from Foramen sphenopalatinum to Fissura orbitalis inferior (roof of Fossa pterygopalatina).
Apertura anterior of Canalis pterygoideus (Vidii) is located at the dorsal margin of Foramen sphenopalatinum. Apertura externa of Canalis rotundus is located lateral to it.

Vessels and nerves of the area of Foramen sphenopalatinum

Fossa pterygopalatina contains the terminal branches of

A. maxillaris

Branches in Fossa pterygopalatina
Aa. palatinae, Rr. alveolares, A. meningea media, branches to Canalis rotundus and Canalis infraorbitalis, a branch to Canalis pterygoideus, A. sphenopalatina

Branches medial to Foramen sphenopalatinum:
A. sphenopalatina is the main branch. It divides into Rr. nasales mediales and laterales

Veins: Plexus pterygoideus

Nerves:

N. maxillaris and N. canalis pterygoidei (a combination of N. petrosus profundus and N. petrosus superficialis)

CAVUM NASI AND FOSSA PTERYGOPALATINA (Figs. 9 to 17)

Fig. 9

Overview

FIG. 9

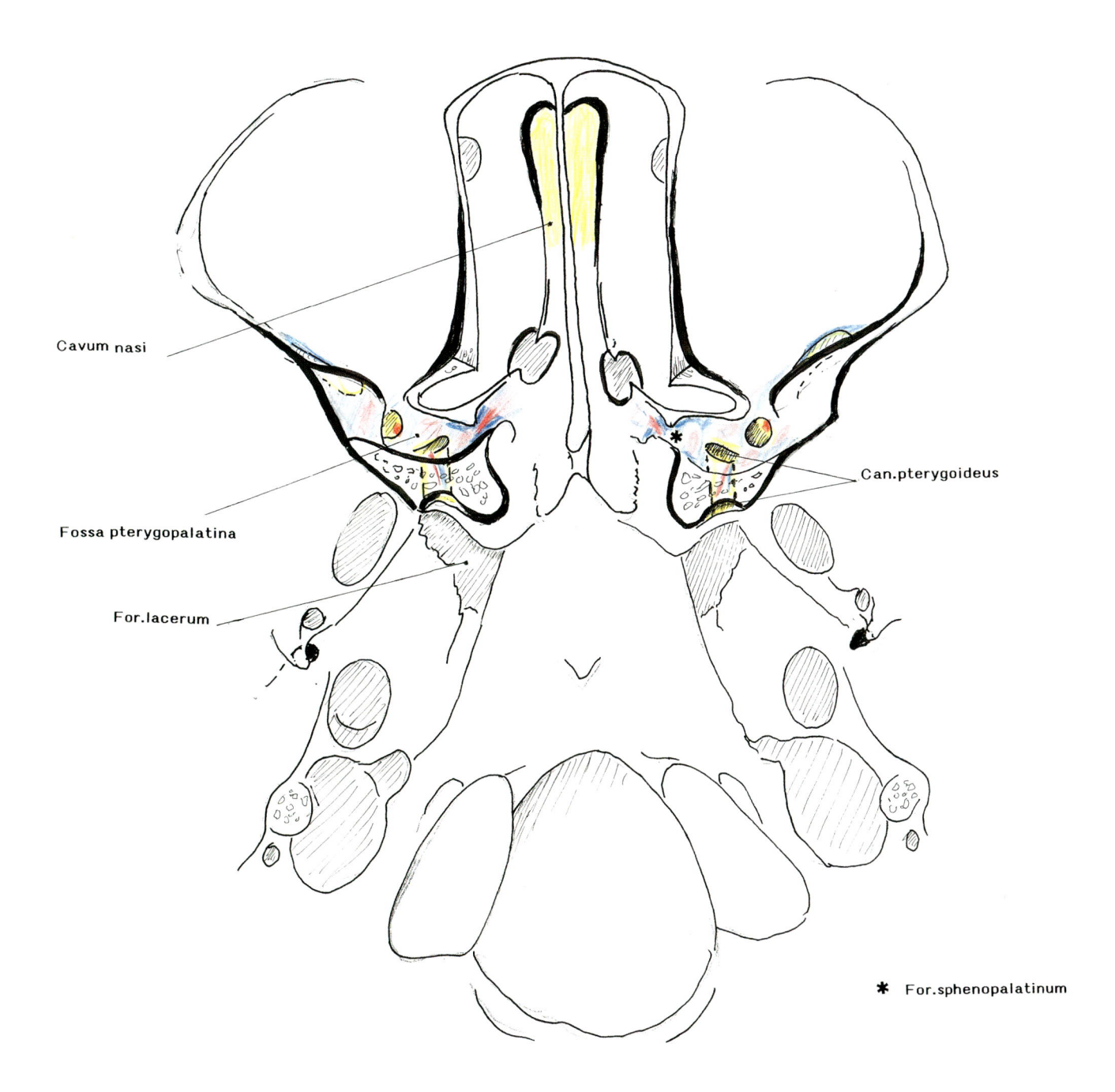

Cavum nasi

Fossa pterygopalatina

For. lacerum

Can. pterygoideus

✱ For. sphenopalatinum

Fig. 10

Wall of Cavum nasi

A Bony structures
B Relief of Mucosa

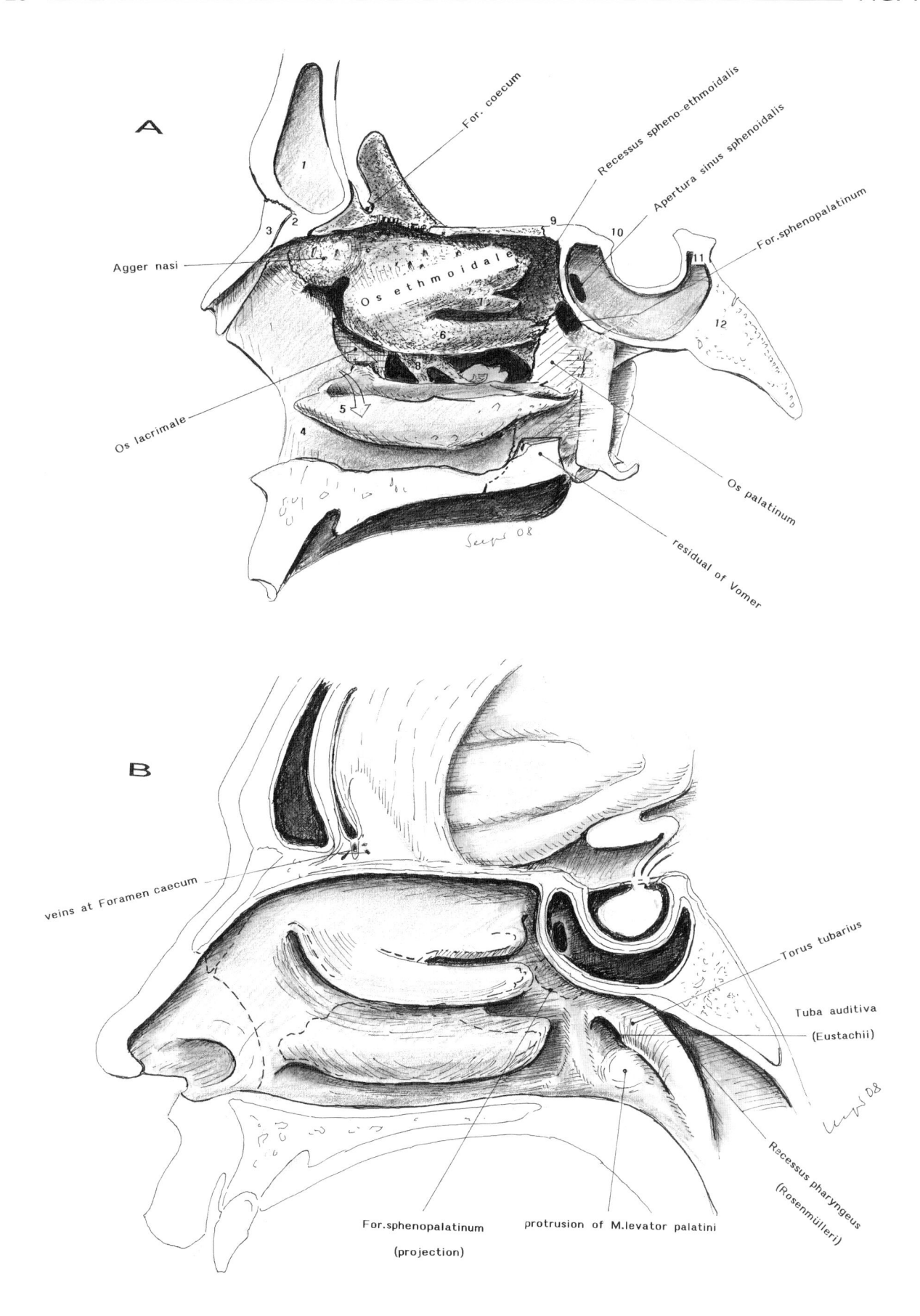

A

For. coecum

Recessus spheno-ethmoidalis

Apertura sinus sphenoidalis

For. sphenopalatinum

Agger nasi

Os ethmoidale

Os lacrimale

Os palatinum

residual of Vomer

B

veins at Foramen caecum

Torus tubarius

Tuba auditiva
(Eustachii)

Recessus pharyngeus
(Rosenmülleri)

For. sphenopalatinum
(projection)

protrusion of M. levator palatini

Fig. 11

Outside of the wall of Cavum nasi, segment of Sinus maxillaris and Orbita

Abbreviations

1 Os nasale
2 Processus front. maxillae
3 Os lacrimale
4 Foramen ethmoidale ant.
5 Lamina perpendicularis
6 Foramen ethmoidale post.
7 Os palatinum, Processus orbitalis
8 Canalis opticus
9 Processus clinoideus ant.
10 Probe in Foramen sphenopalatinum
11 Canalis rotundus
12 Canalis pterygoideus (Vidii)
13 Processus pterygoideus, Lamina lat.
14 as 13, between Ala major (transected) and 10
15 Os palatinum, vertical portion
16 Os palatinum, Processus pyramidalis
16 a) probe in Canalis palatinus major
17 Maxilla, Facies orbitalis
18 Spina trochlearis
19 Corpus sphenoidale, orbital wall of Sinus sphenoidalis
20 Fissura orbitalis sup., medial-basal ground
21 probe in Canalis rotundus (lateral from 22)
22 probe in Canalis pterygoideus (Vidii), see 12 in A
23 Concha media, processus of its: Processus maxillaris
24 Concha media, Processus maxillare
25 Fontanellae
26 Concha media, Processus ethmoidalis
27 Processus uncinatus
28 Hiatus semilunaris
29 insertion of Concha inf. (projection)
30 Ductus nasolacrimalis (projection)
31 Processus uncinatus (as 27)
32 Infundibulum ethmoidale (projection)
33 Hiatus semilunaris as 28
34 Processus uncinatus, connection to 35
35 Concha media, connection to Processus uncinatus
36 Bulla ethmoidalis, bulging into Sinus maxillaris
37 Insertion of Concha media (projection)
38 posterior Fontanella
39 Concha media, Processus maxillaris

FIG. 11

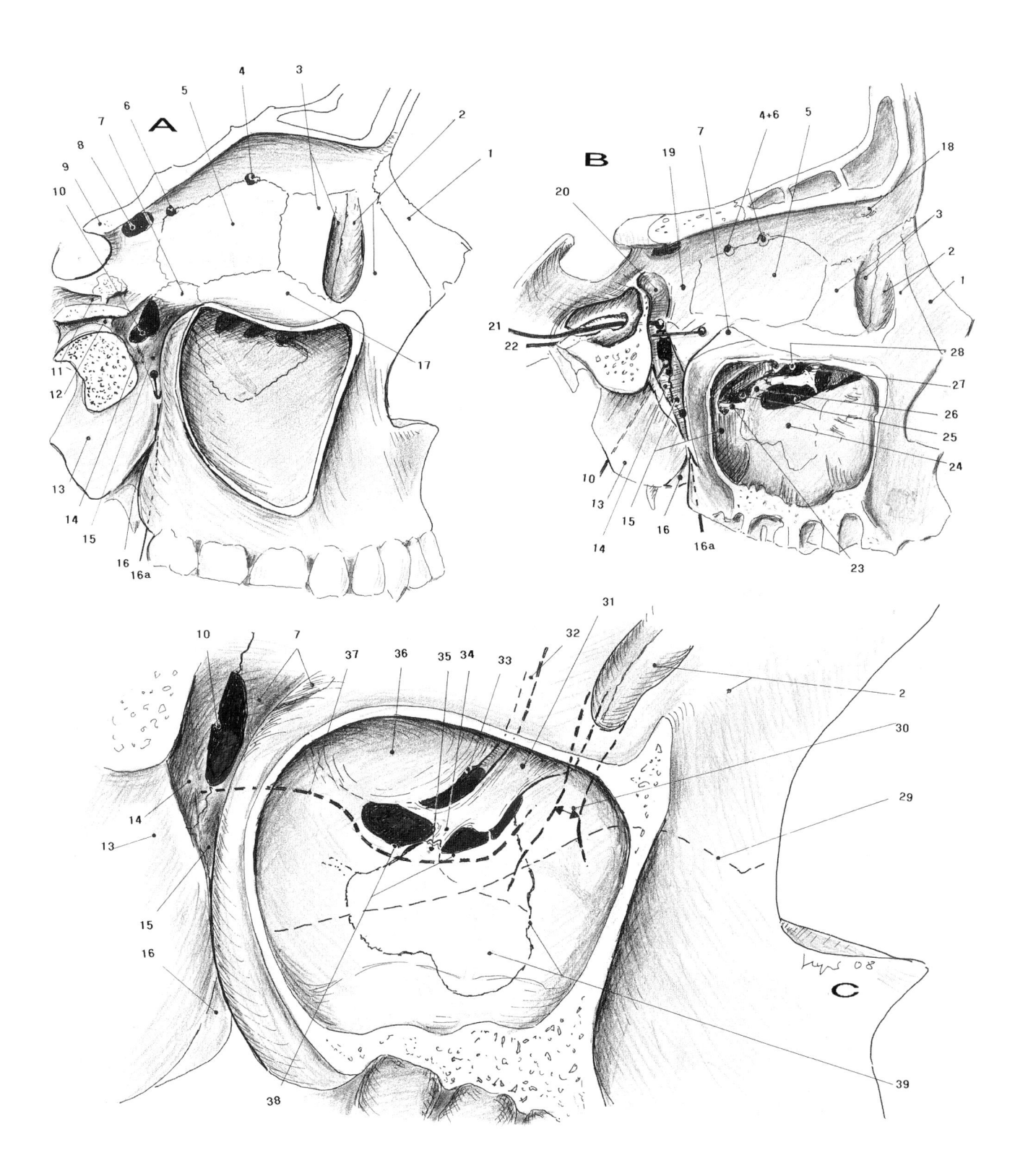

Fig. 12

Connections of parasinuses to Cavum nasi

Abbreviations
1 Processus clinoideus anterior
2 Foramen opticum
3 bulging of Canalis opticus
4 Recessus sphenoethmoidalis of Meatus nasi superior
5 Concha nasalis media (cut)
6 Fossa sacci lacrimalis (projection)

Connections of Sinus paranasales to Meati nasales

(arrows)

a from Sinus sphenoidalis

b from Cellulae ethmoidales postt.

c from Cellulae ethmoidales antt. (here: from Bulla ethmoidalis)

d from Sinus frontalis (here: double), Cellulae ethmoidales antt.,
    and Sinus maxillaris

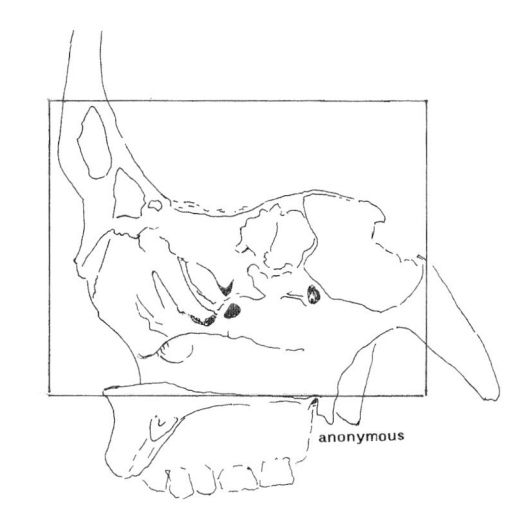

anonymous

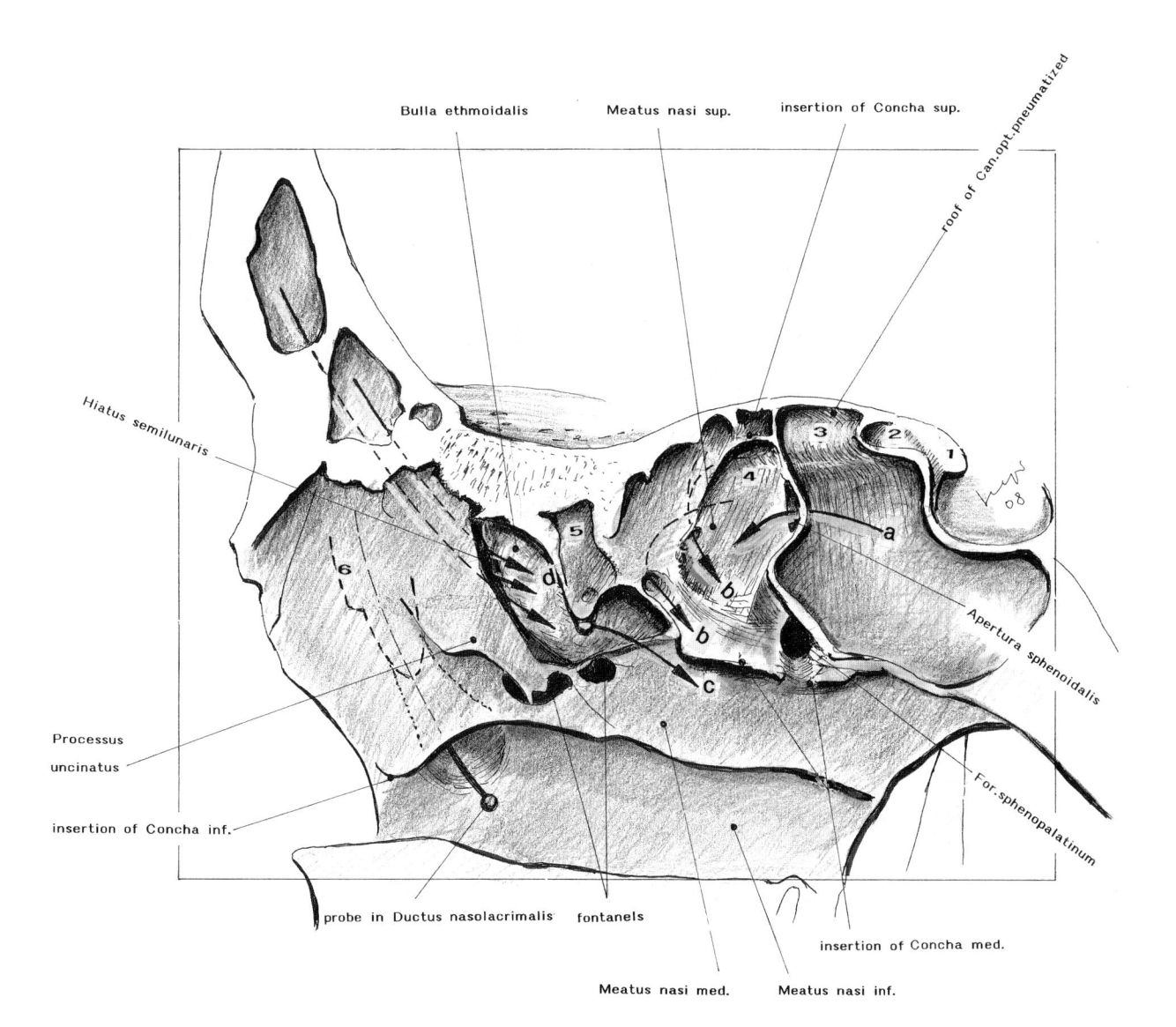

Fig. 13

Choana and surrounding structures

I a: Canales pterygoidei (projections). Course parallel or slightly convergent.
I b: Canales rotundi (projections). Course divergent.
II: Foramen ovale and Canalis rotundus

Abbreviations:
1 Foramen spinosum
2 Ala major
3 Os zygomaticum
4 Foramen palatinum majus
5 Os palatinum, Processus pyramidalis
6 Hamulus pterygoideus
7 Ala vomeris
8 Concha nasalis media
9 bony gap for Synchondrosis sphenooccipitalis (child)

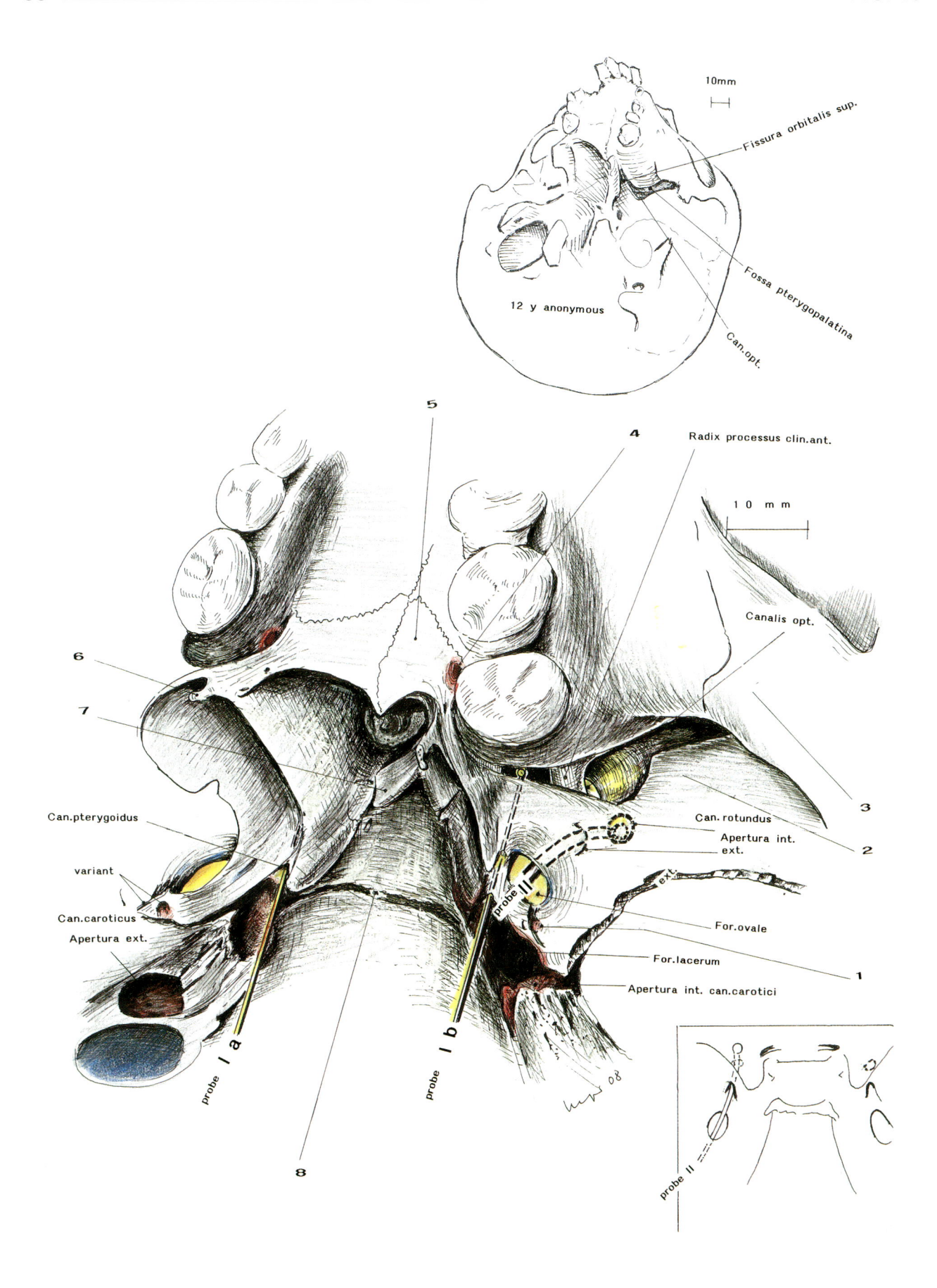

10mm

Fissura orbitalis sup.

Fossa pterygopalatina

Can.opt.

12 y anonymous

5

4

Radix processus clin.ant.

10 mm

Canalis opt.

6

7

3

Can.pterygoidus

Can. rotundus

Apertura int.
ext.

2

variant

Can.caroticus

ext.

Apertura ext.

For.ovale

For.lacerum

probe I a

probe I b

probe III

Apertura int. can.carotici

1

8

probe II

Fig. 14

Continuation of Fig. 13
Os ethmoidale is resected –a-, Os palatinum and Processus pterygoideus are transected
–c- and –d-

a) Lamina papyracea
b) Transection between a and c
c) Transection of Os palatinum, close to Tuber maxillare
d) Transection of the base of Processus pterygoideus

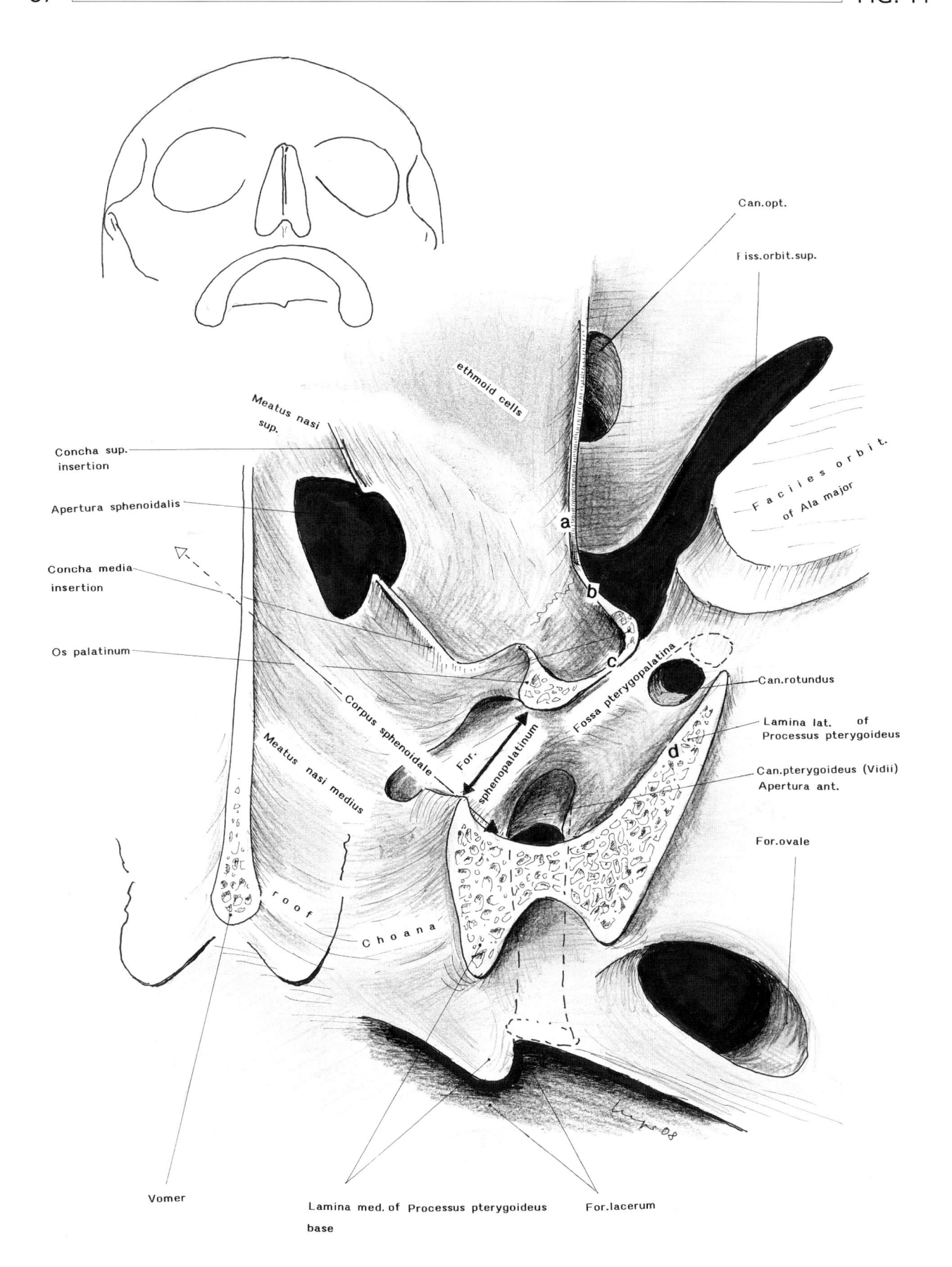

Can.opt.

Fiss.orbit.sup.

ethmoid cells

Meatus nasi sup.

Concha sup. insertion

Apertura sphenoidalis

Concha media insertion

Os palatinum

Corpus sphenoidale

Meatus nasi medius

roof

Choana

a

b

c

Faciies orbit. of Ala major

Fossa pterygopalatina

Can.rotundus

Lamina lat. of Processus pterygoideus

Can.pterygoideus (Vidii) Apertura ant.

For.ovale

For. sphenopalatinum

d

Vomer

Lamina med. of Processus pterygoideus base

For.lacerum

Fig. 15

Fossa pterygopalatina and Foramen sphenopalatinum

A Usual findings
B Variant

Abbreviations
1 Fissura tympanomastoidea
2 Fossa mandibularis, post. segment
3 Porus acusticus ext., anterior wall
4 Os palatinum (Pars orbitalis)
5 Os palatinum, connecting Tuber maxillare and Processus pterygoideus
6 Processus pterygoideus, Lamina lateralis

FIG. 15

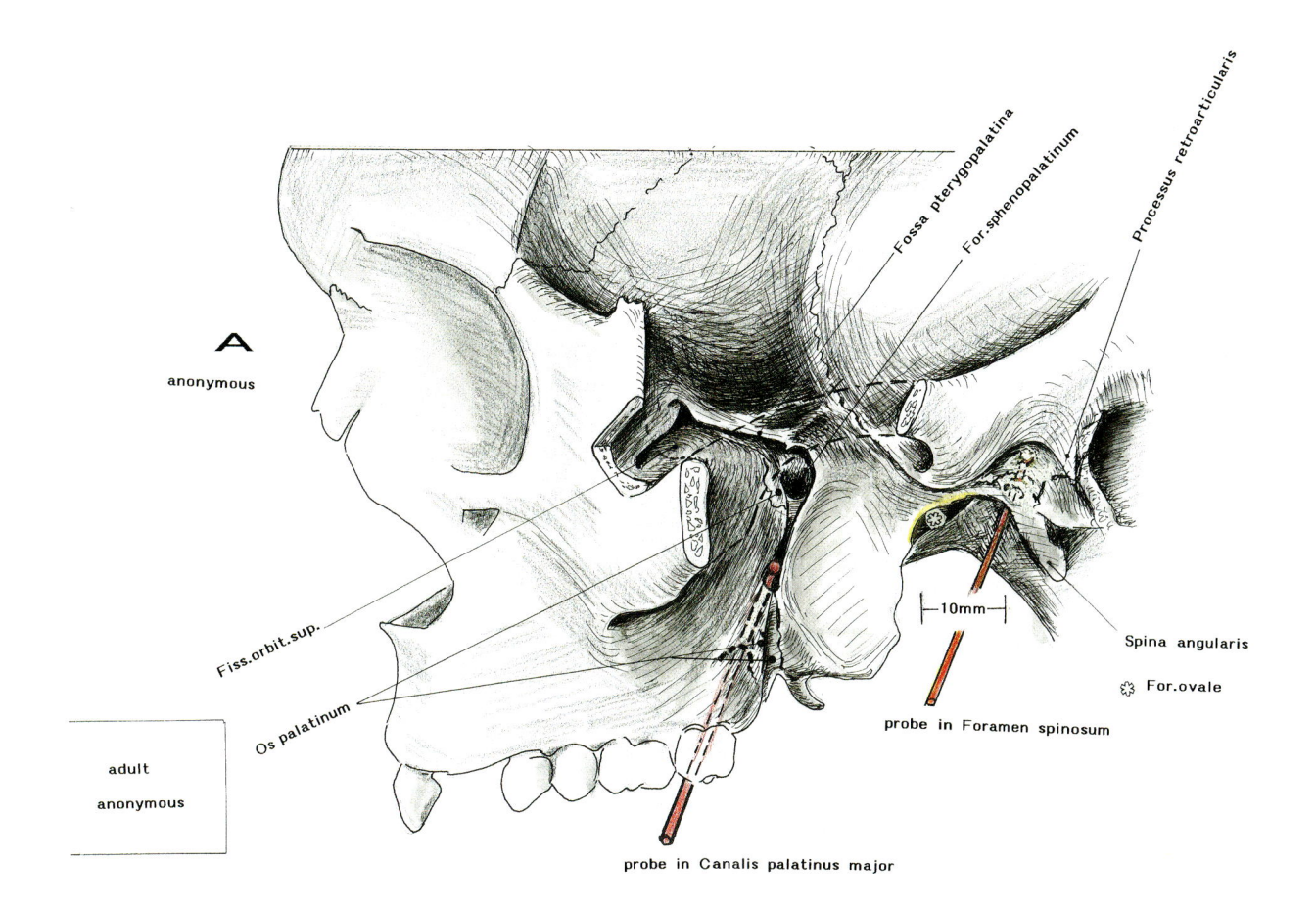

A

anonymous

adult

anonymous

Fossa pterygopalatina

For.sphenopalatinum

Processus retroarticularis

Fiss.orbit.sup.

Os palatinum

Spina angularis

For.ovale

probe in Foramen spinosum

—10mm—

probe in Canalis palatinus major

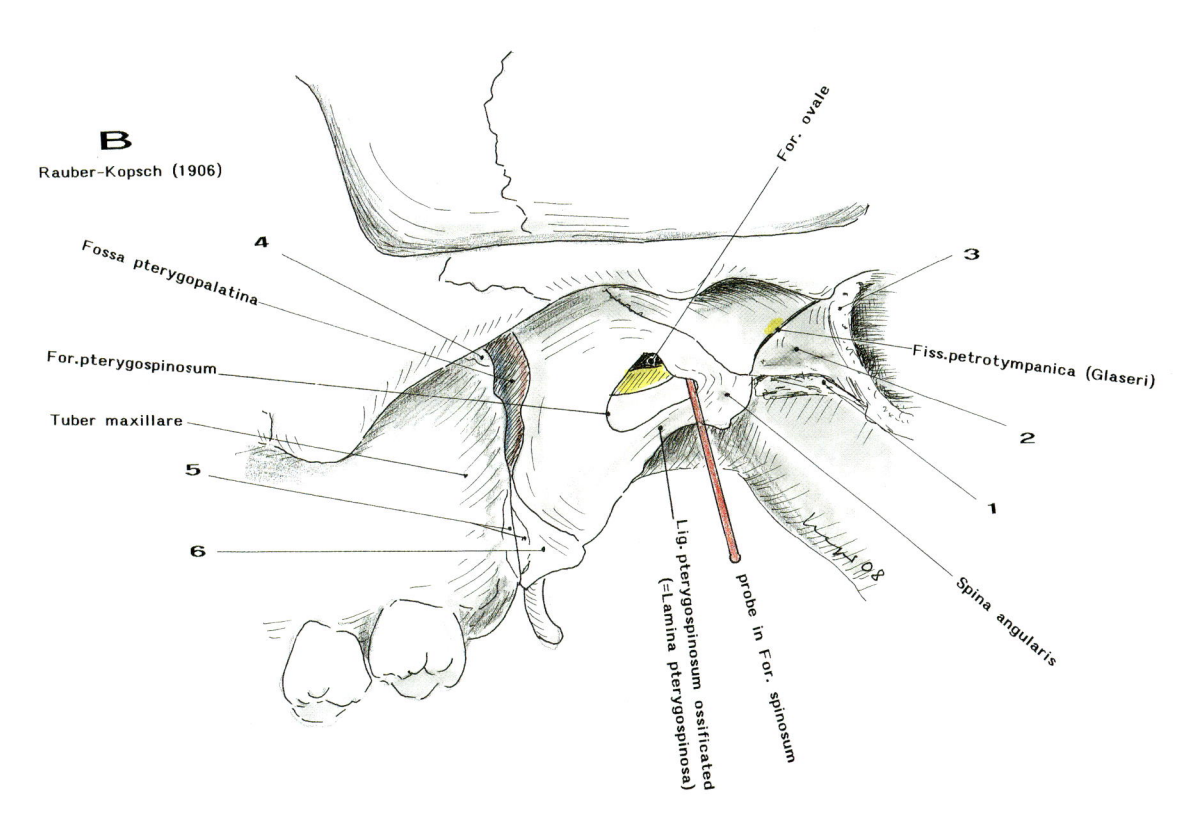

B

Rauber-Kopsch (1906)

For. ovale

Fossa pterygopalatina

For.pterygospinosum

Tuber maxillare

4

3

Fiss.petrotympanica (Glaseri)

2

5

6

1

Lig.pterygospinosum ossificated
(=Lamina pterygospinosa)

probe in For. spinosum

Spina angularis

Fig. 16

Fissura orbitalis inf. (and sup.) and Fossa pterygopalatina

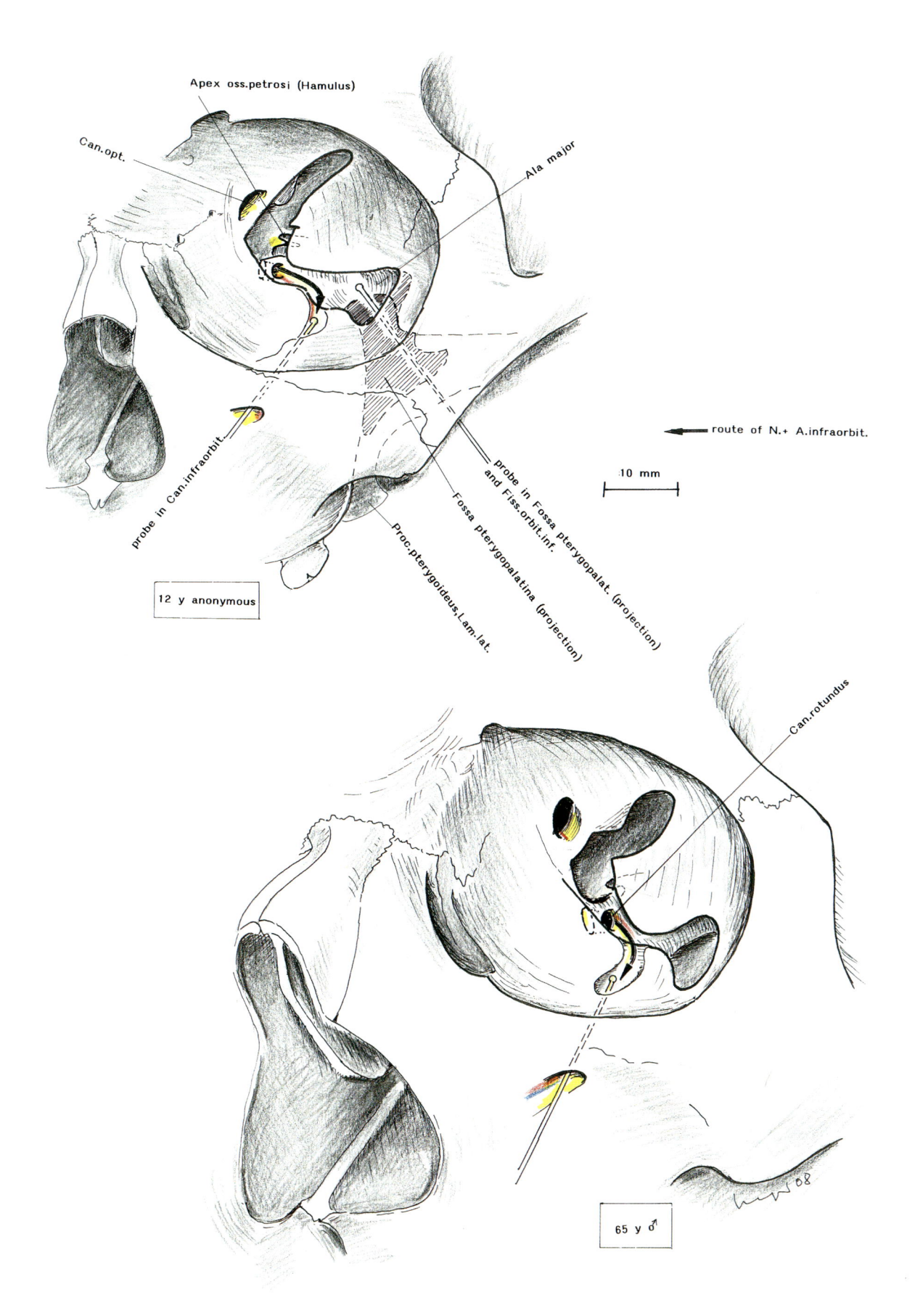

Apex oss.petrosi (Hamulus)

Can.opt.

Ala major

route of N.+ A.infraorbit.

10 mm

probe in Can.infraorbit.

probe in Fossa pterygopalat. (projection)
and Fiss.orbit.inf.

Fossa pterygopalatina (projection)

Proc.pterygoideus.Lam.lat.

12 y anonymous

Can.rotundus

65 y ♂

Fig. 17

Fissura orbitalis inferior and Fossa pterygopalatina

Probe I: Canalis palatinus major (A.+V. palatinus/a major and N. palatinus ant.)
Probe II: Canalis pterygoideus Vidii (N. pterygoideus and A.+V. pterygoideus/a)
Arrow: Course of N. maxillaris (A. + V. infraorbitalis).

Abbreviations
1   Processus pterygoideus, Lamina lateralis
2   Fossa scaphoidea (adherent to the distal segment of Tuba)
3   Sutura squamosa
4   Ala major, Facies temporalis
5   Os palatinum
6   Os palatinum, Pars verticalis-Pars horizontalis (Processus pyramidalis), transition
7   Tuber maxillare

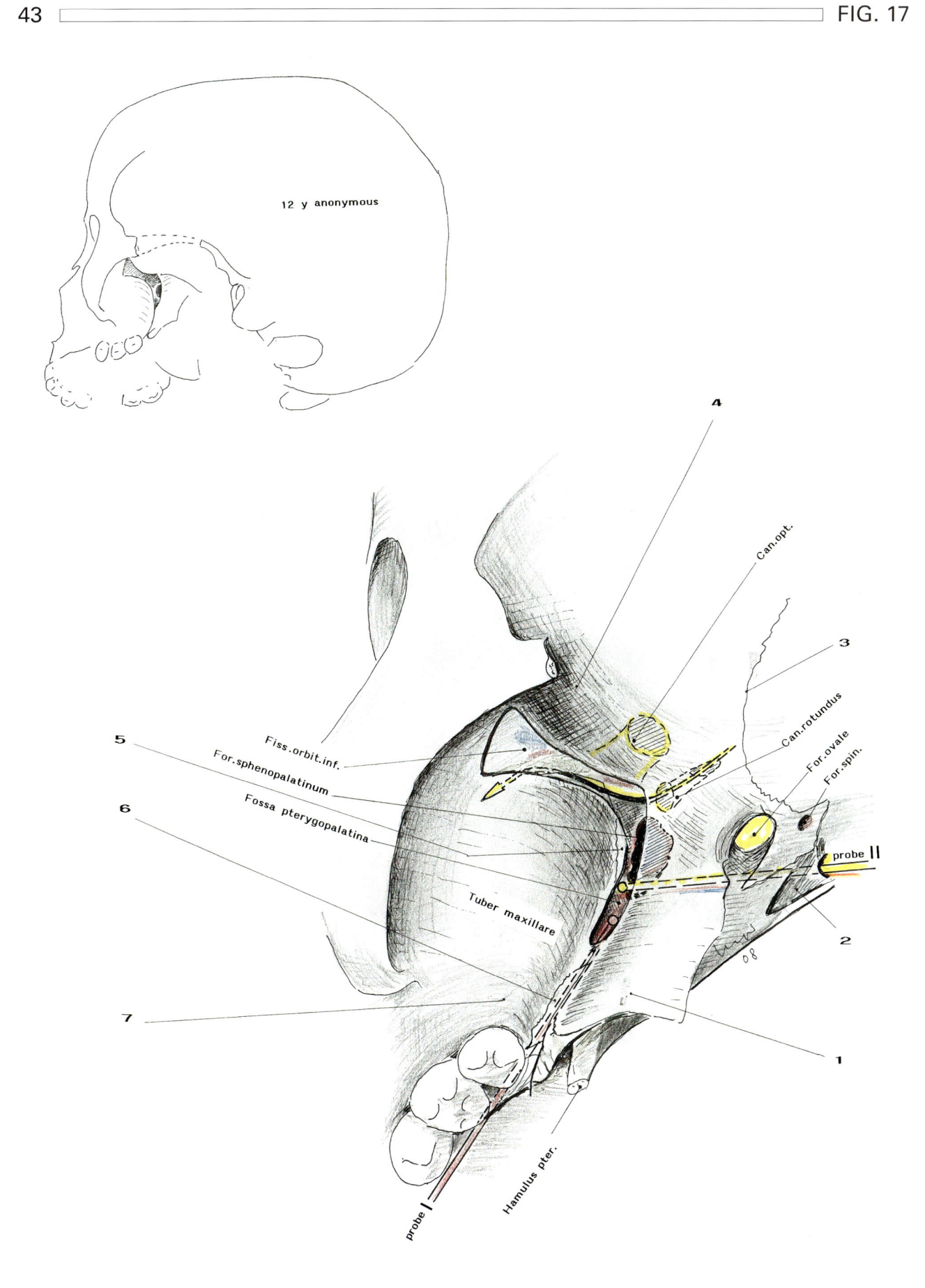

12 y anonymous

Can.opt.

4

3

Can.rotundus

For.ovale

For.spin.

Fiss.orbit.inf.

5

For.sphenopalatinum

Fossa pterygopalatina

6

probe II

Tuber maxillare

2

08

7

1

probe I

Hamulus pter.

Literature

Castelnuovo P, Delfre G, Locatelli D, Padoan G, de Bernardi F, Pistochini A, Bignami M (2006) Endonasal endoscopic duraplasty. Our experience. Skull base 16 (1), 15–8

Castelnuovo P, Locatelli D (2007) The endoscopic surgical technique. "Two nostrils – four hands". Endo-PressTM, Tuttlingen, Germany

Castelnuovo P, Locatelli D, Mauri S, de Bernadi F (2003) Extended endoscopic approaches to the skull base, anterior cranial base CSF leaks. In: De Divilis E, Cappabianca P (eds): Endoscopic endonasal trans-sphenoidal surgery. Springer, Wien New York, pp 137–138

Castelnuovo P, Locatelli D, Santi L, Emanuelli E, Pagella F, Canevari FR (1998) Sinonasal endoscopic accesss to the pituitary gland. In: Stamberger, Wolf eds. ERS & ISIAN Meeting, Monduzzi, pp 337–339

Castelnuovo P, Mauri S, Locatelli D, Emanuelli E, Delu G, Giulio G (2001) Endoscopic repair of cerebrospinal fluid rhinoliquorrhea. Learning from our failures. Am J Rhinol 15, 333–342

Cavallo LM, Messina A, Cappabianca P, Esposito F, de Divitis E, Gardner P, Tschabitscher M (2005) Endoscopic endonasal surgery of the midline skull base.

Anatomical study and clinical considerations. Neurosurg Focus 19 (1), E2, pp 1–14

De Divitis E, Cappabianca P, Cavallo LM (2002) Endoscopic transethmoidal approach: adaptility of the procedure to different sellar lesions. Neurosurgery 51 (3), pp 699–705

Draf W (1978) Die Endoskopie der Nasenhöhlen. Springer, Berlin Heidelberg New York

Draf W (1982) Die chirurgische Behandlung entzündlicher Erkrankungen der Nasennebenhöhlen. Indikation, Operationsverfahren, Gefahren. Fehler und Komplikationen, Revisionschirurgie. Arch Otorhinolaryngol 235, 133–305

Draf W (1983) Endoscopy of the paranasal sinuses. Springer, Berlin Heidelberg New York

Grisoli F, Vincentelli F, Henry J (1982) Anatomical bases for the transsphenoidal approach to the pituitary gland. Anat Chir 3, 207–220

Iho HD (2001) Endoscopic transsphenoidal surgery. J Neurooncol 54 (2) pp 187–195

Jovanovic S (1961) Supernumerary frontal sinuses on the roof of the orbit: Their clinical significance. Acta Anat (Basel) 45, 133

Kassam A, Snyderman CH, Mintz A, Gardner P, Carrau RL (2005) Expanded endonasal approach: The rostrocaudal axis. Part II. Posterior clinoides to the foramen magnum. Neurosurg Focus Jul 15; 19 (1): E4

Kassam AB, Gardner P, Snyderman C, Mintz A, Carrau R (2005) Expended endonasal approach: Fully endoscopic, completely transnasal approach to the middle third of the clivus, petrous bone, middle cranial fossa, and infratemporal fossa. Neurosurg Focus Jul; 15:19(1):E6

Keros P (1965) Über die praktische Bedeutung der Niveauunterschiede der Lamina cribrosa des Ethmoids. Z Laryngol Rhinol 41, 808

Krmpotic-Nemanic J (1977) Entwicklungsgeschichte und Anatomie der Nase und der Nasennebenhöhlen in Hals-Nasen-Ohrenheilkunde in Praxis und Klinik. In: Berendes-Link-Zöllner (1977), Bd 1, Obere und untere Luftwege. Thieme, Stuttgart

Lang J (1979) Kopf, Teil B, Gehirn- und Augenschädel. Springer, Berlin Heidelberg New York

Lang J (1981) Neuroanatomie der Nn. opticus, trigeminus, facialis, glossopharyngeus, vagus, accessorius und hypoglossus. Arch. Otorhinolaryngol 231, 1–69

Lang J, Sacals LE (1981) Über die Höhe der Cavitas nasi, die Länge ihres Bodens und Masse, sowie Anordnung der Conchae nasales und der Apertura sinus sphenoidalis. Anat Anz 149, 297–318

Locatelli D, Rampa F, Acchiardi I, Bignami M, de Bernardi F, Castelnuovo P (2006) Endoscopic endonasal approaches for repair of cerebrospinal fluid leaks. Nine-year experience. Neurosurgery 58 (Suppl 2), ONS 246–256

Stamm WK (1981) Anatomy of the pterygopalatine foramen and the Fontanella in the lateral nasal wall. Rhinology 19, 87–91

CHAPTER III
SINUS SPHENOIDALIS AND
FOSSA PTERYGOPALATINA
(Figs. 18 to 34)

Overview (Figs. 18 to 20)

Sinus sphenoidalis is enclosed by Corpus sphenoidale. Corpus sphenoidale and Pars basilaris of the occipital bone are merged with Os basilare in adults.
In children and adolescents and many mammals, both segments of Os basilare are connected by Synchondrosis sphenooccipitalis.

Sinus sphenoidalis (Figs. 19 and 20)

It is enclosed by Corpus sphenoidale, which is merged with the adjacent segments of the sphenoid bone since the perinatal period of life. Outside structures cause inward bulging of the walls of the sinus (Fig. 20), Fig. 20 presents a widening of the sinus. A bulging of Canalis opticus and of the siphon area of the carotid artery can be observed easily. The optocarotid recess divides both structures. The walls of Canalis rotundus and Canalis pterygoideus can be penetrated by a fine needle, if the wall of the sinus is very thin-walled. The relief of the sinus is better recognized, if a light source is positioned behind the anatomical dissection.

Widening of adjacent structures (Figs. 27, 28, and 33).

Widenings of the sinus to Orbita were presented before, and in Fig. 33, further widenings in Fig. 33. Rare widenings include pneumatizations of Ala major, far lateral from Canalis rotundus. The roof of Canalis opticus may be doubled by a connection of the sinus to a pneumatized Processus clinoideus anterior. It may be combined with a pneumatization of the root of Processus clinoideus ant., which connects the sinus to the clinoid process inferior to Canalis opticus.

Area between Sinus sphenoidalis and Foramen lacerum (Figs. 19 to 34)

Anterior wall of Sinus sphenoidalis

Apertura sinus sphenoidalis is interposed between the insertions of Concha superior and Concha media (Fig. 30). Recessus sphenoethmoidalis (roof of Meatus nasi superior) is located superior to Apertura sphenoidalis.

Foramen sphenopalatinum (Figs. 23 to 25)

Its upper margin is the flat shape of the anterobasal wall of Sinus sphenoidalis at its transition to the roof of Fossa pterygopalatina (* in Fig. 23). Foramen sphenopalatinum is interposed between the vertical segment of Os palatinum and the base of Processus pterygoideus Figs. 23 to 25)

Fossa pterygopalatina (Figs. 22 to 26, and 28 to 33)

The central area of its roof represents Apertura externa of Canalis rotundus, between the base of the insertion of Os palatinum and the base of Processus pterygoideus, lateral to Apertura ant. of Canalis pterygoideus (see probe and arrow in Fig. 23, and Fig. 30).

Canalis pterygoideus penetrates the base of Processus pterygoideus. It connects the roof of Fossa pterygopalatina to the anterior shape of Foramen lacerum.

At the level of the anterior Apertura of Canalis pterygoideus, A. carotis int. is located at the inside of the skull base, at the coronal level of Dorsum sellae. This area is located close to the lumen of the sinus. It is defined after widening of the well known approaches to Sella. This area is illustrated by the cadaver skull dissection in Fig. 28 (see Sulcus caroticus, colored). Drilling of the base of Processus pterygoideus along Canalis pterygoideus reveals the carotid artery between its siphon and Foramen lacerum, close to Apex pyramidis. Apex encloses Apertura int. of Canalis caroticus.

Foramen lacerum and its contents (Figs. 23 to 31)

Foramen lacerum lies between the base of Processus pterygoideus, Ala major, Apex pyramidis (enclosing Apertura int. of the carotid channel) and the bony bloc of Os basilare, at the level of Synostosis (Synchondrosis) sphenooccipitalis. Its extra- and intracranial shapes are incongruent and variable (Figs.1 to 7). A carotis int. fills Foramen lacerum. It is covered by chondroid layers and ligaments. No bony structure is interposed between the external and internal shape of the cranial base. Between the anterior margin of Foramen lacerum and the posterior clinoid process, the artery bends in a steep course, distant to the external shape of the cranial base, along the wall of Sinus sphenoidalis (Fig. 21).

SINUS SPHENOIDALIS AND FOSSA PTERYGOPALATINA (Figs. 18 to 34)

Fig. 18

Overview

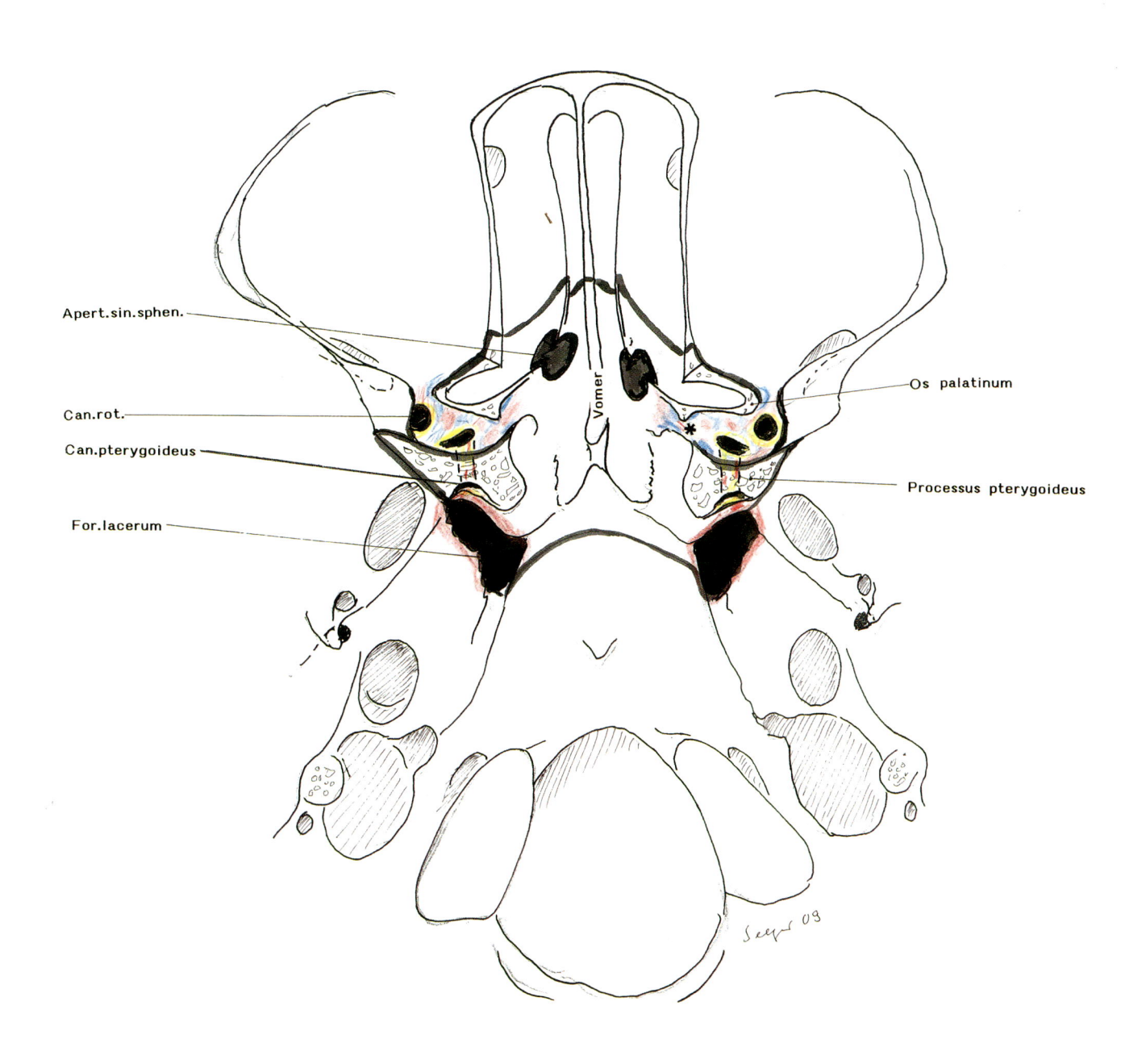

Apert.sin.sphen.

Can.rot.

Can.pterygoideus

For.lacerum

Vomer

Os palatinum

Processus pterygoideus

Fig. 19

Os sphenoidale

A. According to Rauber-Kopfsch (1907), anterior view direction
B. As A, posterior view direction
A' Similar to A, cadaver skull dissection, sectional enlargement
B' Similar to B, cadaver skull dissection, sectional enlargement
A'' and **B''** anatomical sketches for topograms

Abbreviations
1 Crista sphenoidalis (Alae vomeris removed)
1' Alae vomeris (projection)
2 anterior wall of Sinus sphenoidalis
3 Apertura sinus sphenoidalis
4 Canalis pterygoideus Vidii, Apertura ant.
4' Canalis pterygoideus Vidii, Apertura post.
5 Canalis rotundus, Apertura ant.
5' Canalis rotundus, Apertura post.
6 Ala major, Facies infratemporalis
7 Ala major, Facies orbitalis
8 Fissura orbitalis superior
9 Planum sphenoidale
10 Fossa scaphoidea
11 Ala major, Facies cerebralis
12 Sulcus caroticus
13 Dorsum sellae
14 Corpus sphenoidale
15 Fossa cranii media (projection)
16 Ala major, Facies infratemporalis
17 Processus pterygoideus, Lamina medialis

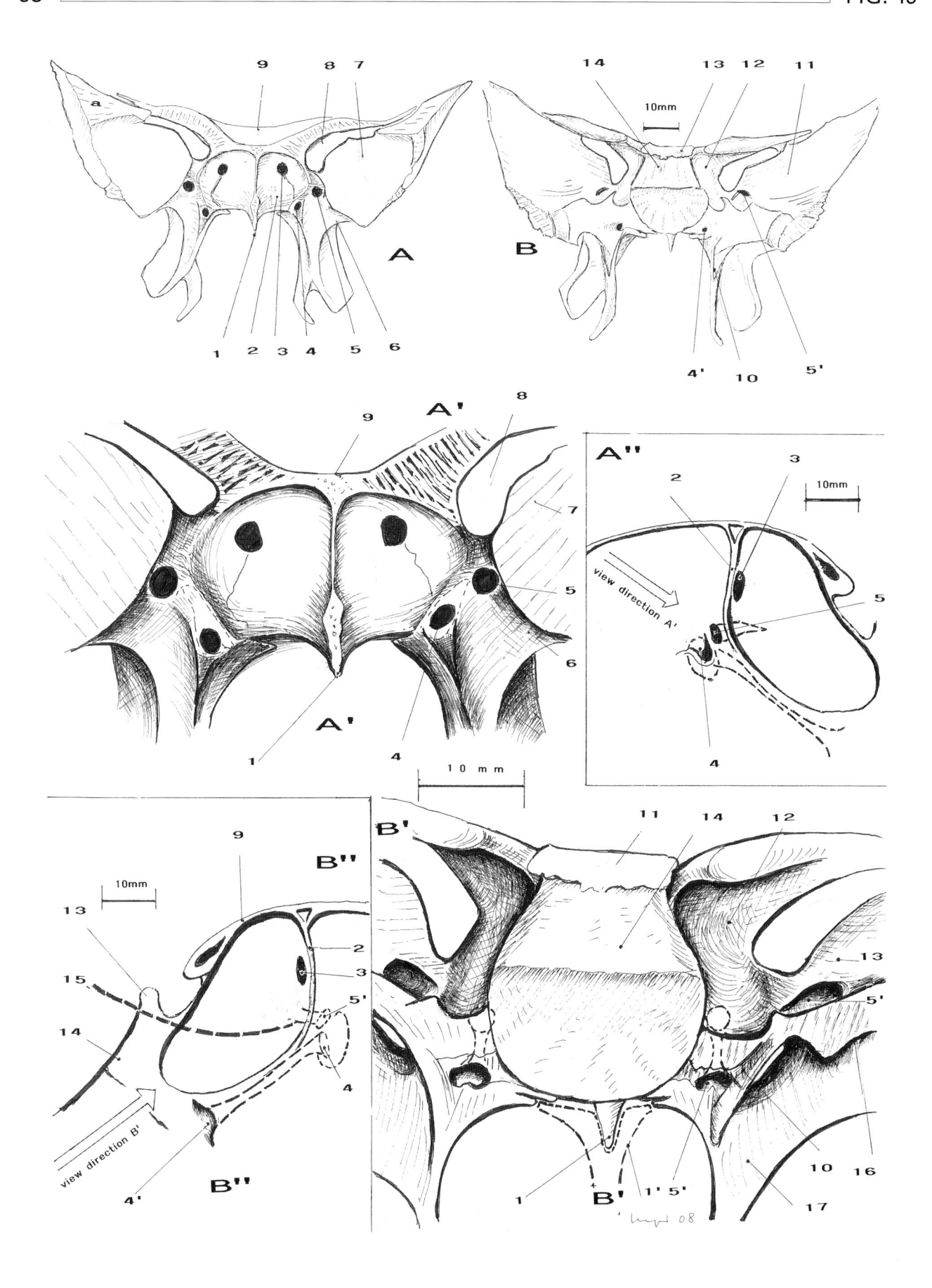

Fig. 20

Sinus sphenoidalis, wall

Thin walled sinus (common variant). The bulging of the wall by A. carotis int. is accentuated. Walls of Canalis rotundus and Canalis pterygoideus removed.

Abbreviations
1 Meatus nasi sup.
2 Concha sup.
3 atypical Cella ethmoidalis between Apertura sinus sphenoidalis and Planum sphenoidale
4 atypical supraoptic widening of Sinus sphenoidalis
5 optic nerve prominence (Divitis et al, 2006)
6 carotid protuberance
7 Tuberculum sellae
8 Processus clinoideus ant.
9 Curvatura post. of A. carotis int. (here: prominent to Sinus sphenoidalis)
10 Foramen lacerum (may be located close to the posterior-lateral wall of the sinus)
11 Apertura int. of Canalis caroticus
12 Apertura posterior of Canalis pterygoideus (Vidii)
13 Lamina med. of Processus pterygoideus
14 Lamina lat. of Processus pterygoideus
15 insertion of Concha nas. inf.
16 root of Ala major
17 Apertura ext. of Canalis rotundus
17a Apertura int. of Canalis rotundus
18 insertion of Concha nas. med.
19 Apertura ant. of Canalis pterygoideus (Vidii)
20 Canalis rotundus, opened

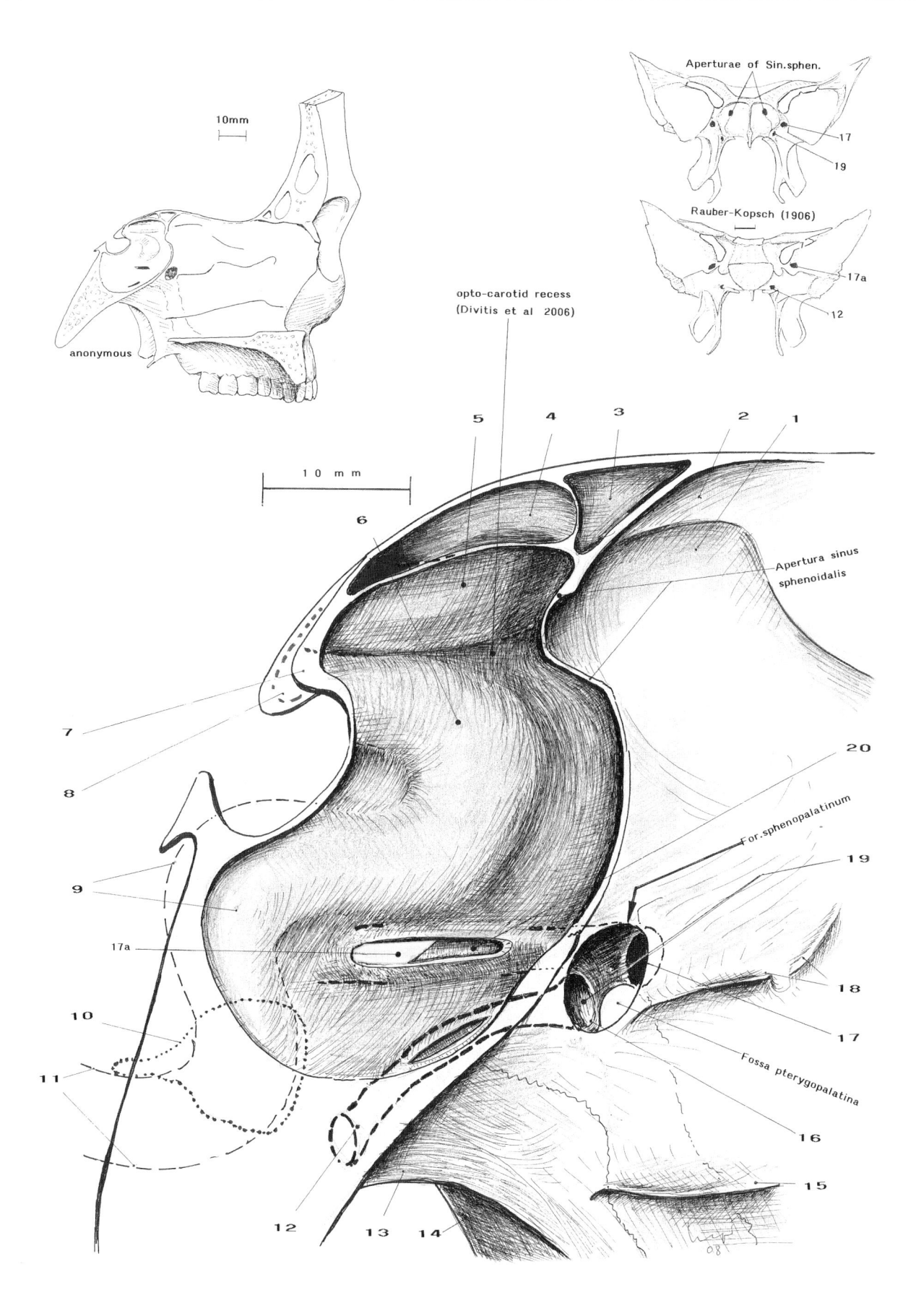

opto-carotid recess
(Divitis et al 2006)

Aperturae of Sin.sphen.

Rauber-Kopsch (1906)

anonymous

10mm

10 m m

Apertura sinus
sphenoidalis

For.sphenopalatinum

Fossa pterygopalatina

Fig. 21

Addendum for Fig. 20
Cranial nerves and blood vessels

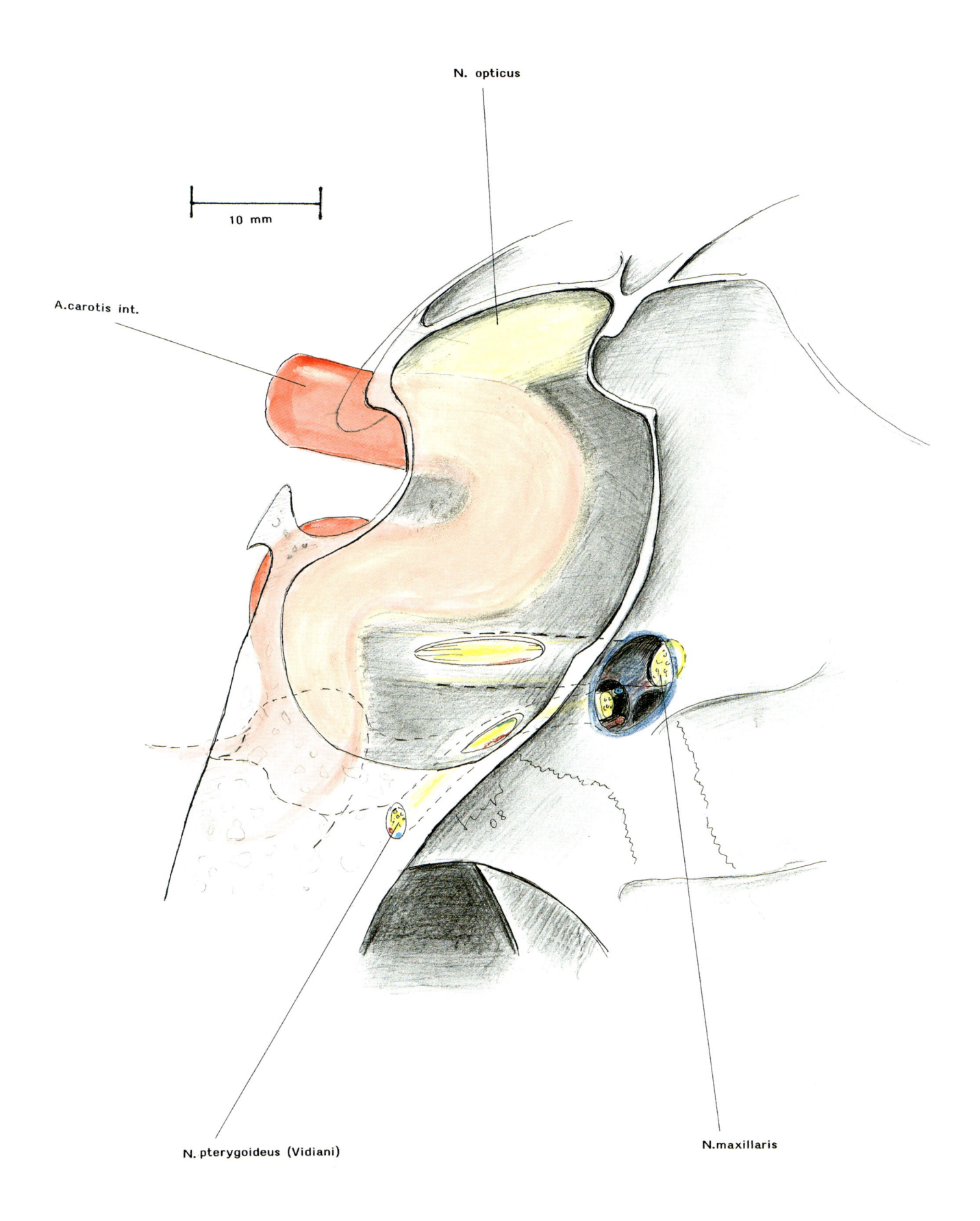

N. opticus

A.carotis int.

10 mm

N. pterygoideus (Vidiani)

N.maxillaris

Fig. 22

Sinus sphenoidalis and roof of Foramen sphenopalatinum. Overview

FIG. 22

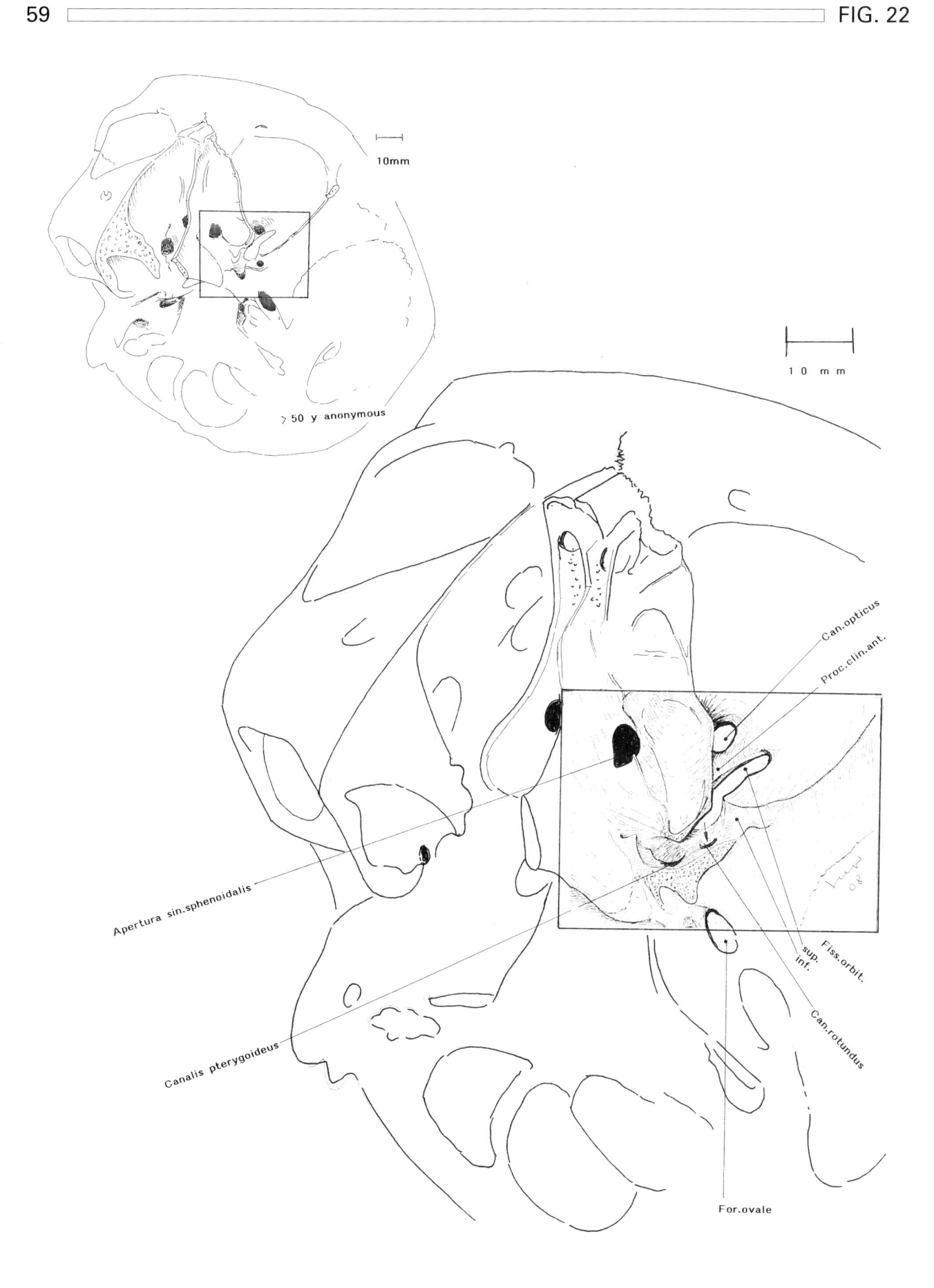

10mm

> 50 y anonymous

10 mm

Can.opticus

Proc.clin.ant.

Apertura sin.sphenoidalis

Fiss.orbit.
sup.
inf.

Can.rotundus

Canalis pterygoideus

For.ovale

Fig. 23

Continuation of Fig. 22. Sectional enlargement

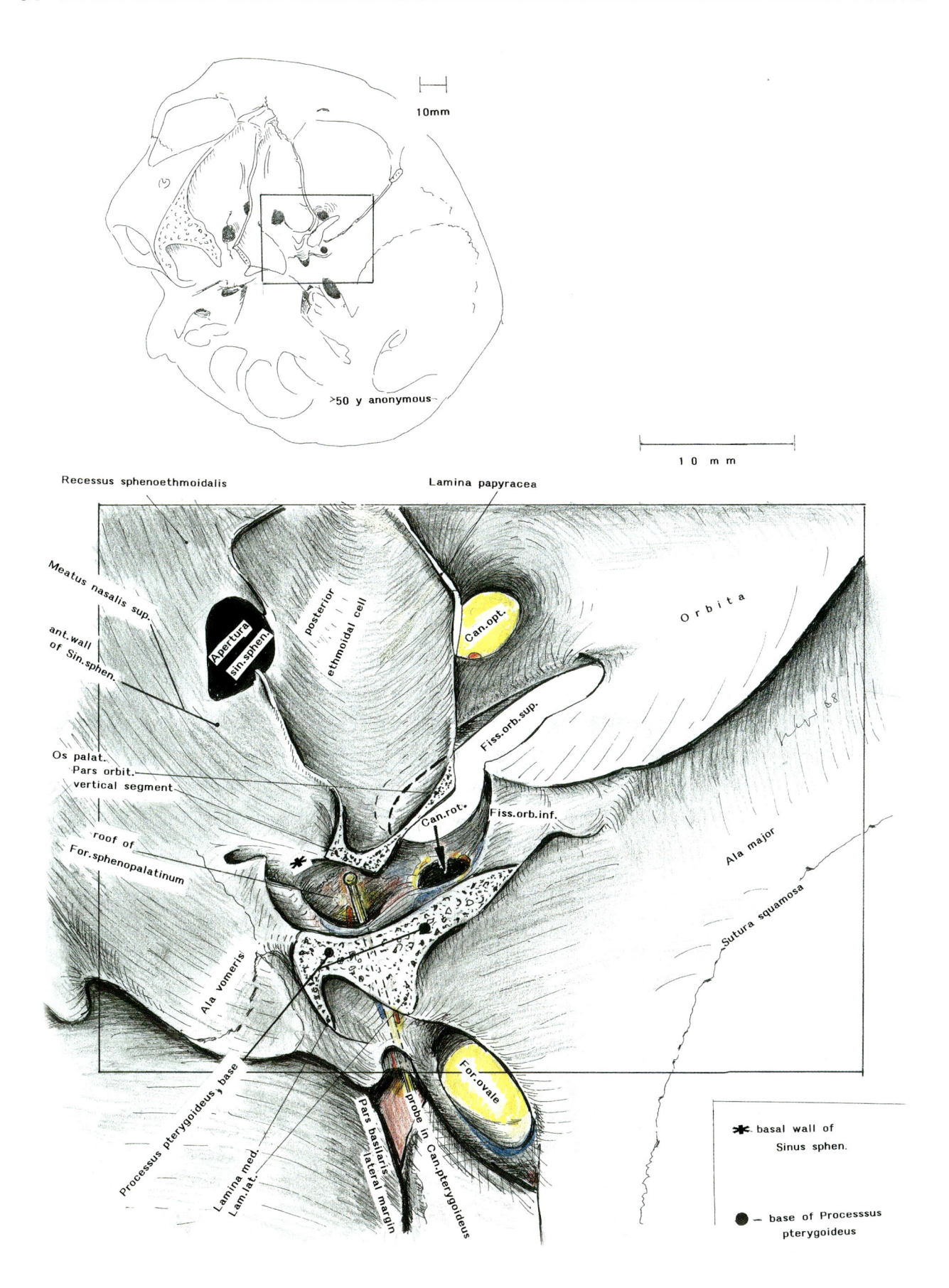

10mm

>50 y anonymous

10 mm

Recessus sphenoethmoidalis

Lamina papyracea

Meatus nasalis sup.

ant.wall
of Sin.sphen.

Apertura
sin.sphen.

posterior
ethmoidal cell

Can.opt.

Orbita

Fiss.orb.sup.

Os palat.
Pars orbit.
vertical segment

Can.rot.

Fiss.orb.inf.

roof of
For.sphenopalatinum

Ala major

Ala vomeris

Sutura squamosa

Processus pterygoideus, base

Lamina med.
Lam.lat.

Pars basilaris,
lateral margin

probe in Can.pterygoideus

For.ovale

✱ basal wall of
Sinus sphen.

● – base of Processus
pterygoideus

Fig. 24

After resection of the left anterior wall of Sinus sphenoidalis

FIG. 24

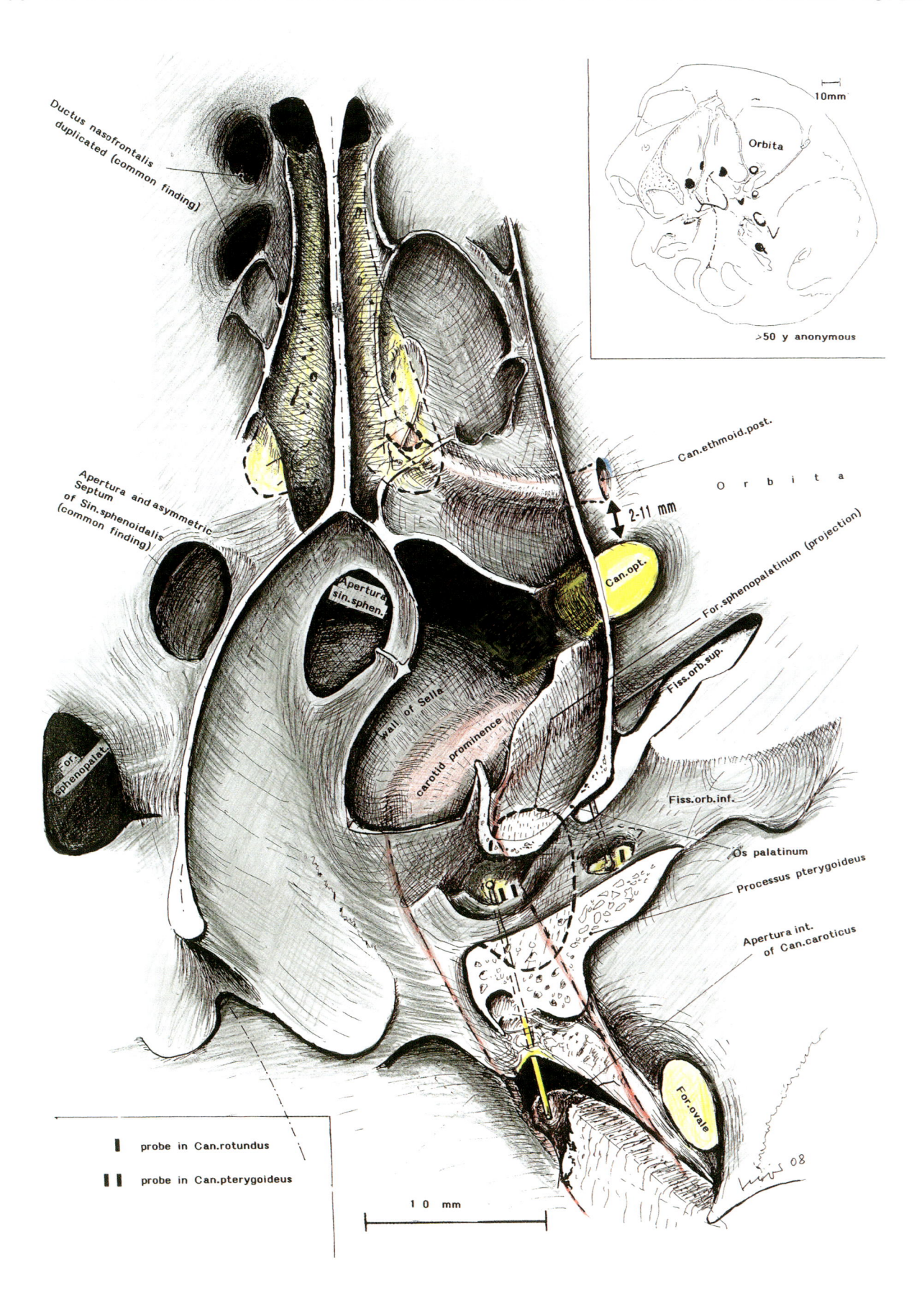

Ductus nasofrontalis duplicated (common finding)

Apertura and asymmetric Septum of Sin.sphenoidalis (common finding)

Apertura sin.sphen.

wall of Sella

carotid prominence

For. sphenopalat.

Can.ethmoid.post.

Orbita

2-11 mm

Can.opt.

Orbita

For.sphenopalatinum (projection)

Fiss.orb.sup.

Fiss.orb.inf.

Os palatinum

Processus pterygoideus

Apertura int. of Can.caroticus

For. ovale

10mm

>50 y anonymous

I   probe in Can.rotundus

II probe in Can.pterygoideus

1 0   mm

Lewis 08

Fig. 25

Resection continued. Sectional enlargement

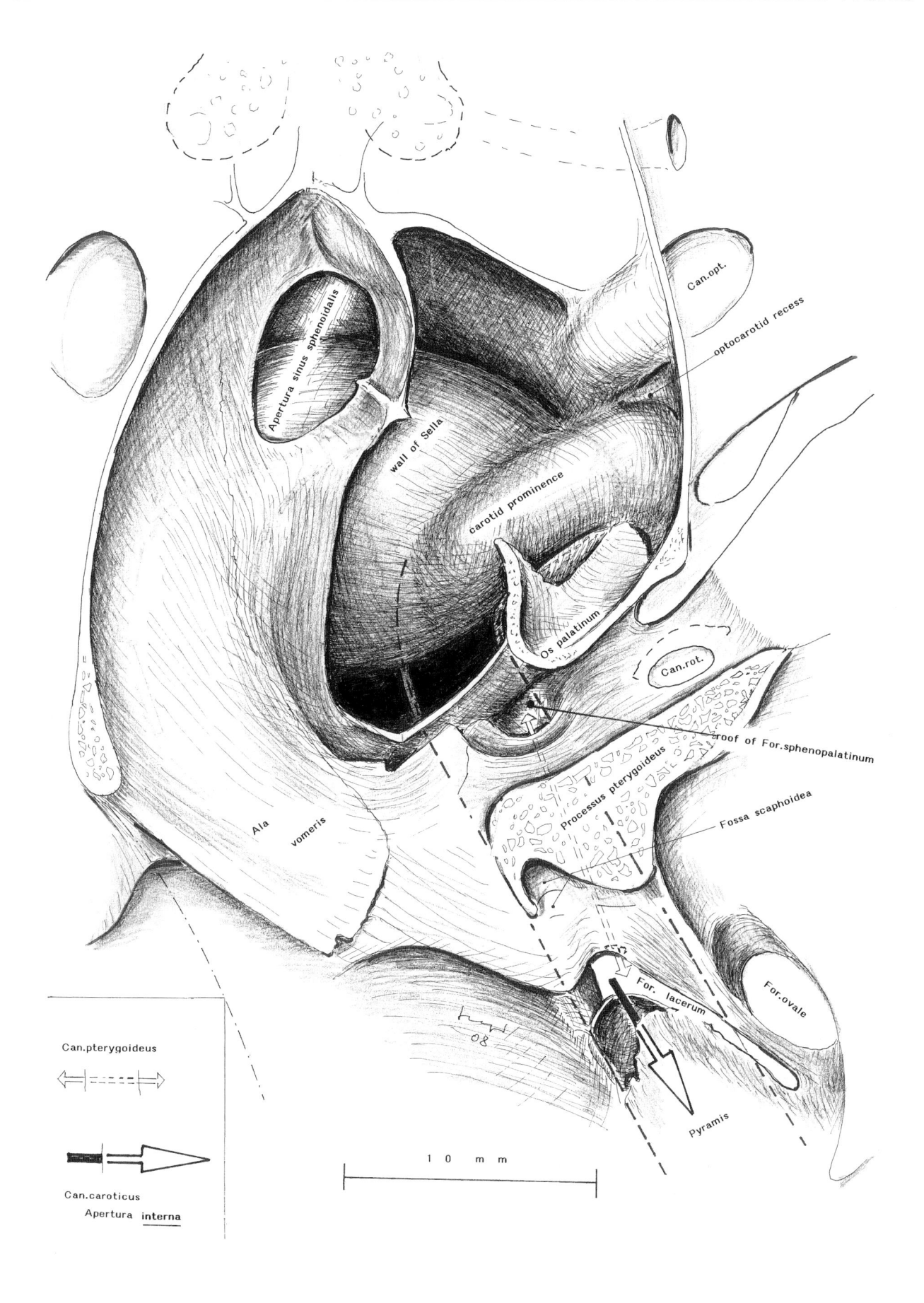

Apertura sinus sphenoidalis

wall of Sella

carotid prominence

Can.opt.

optocarotid recess

Os palatinum

Can.rot.

roof of For.sphenopalatinum

Processus pterygoideus

Fossa scaphoidea

Ala vomeris

For. lacerum

For.ovale

Pyramis

08

Can.pterygoideus

Can.caroticus
Apertura interna

10 mm

Fig. 26

Addendum for Fig. 25
Note close distance between A. (and V.) ethmoidalis post. and Canalis and N. opticus
(common finding)

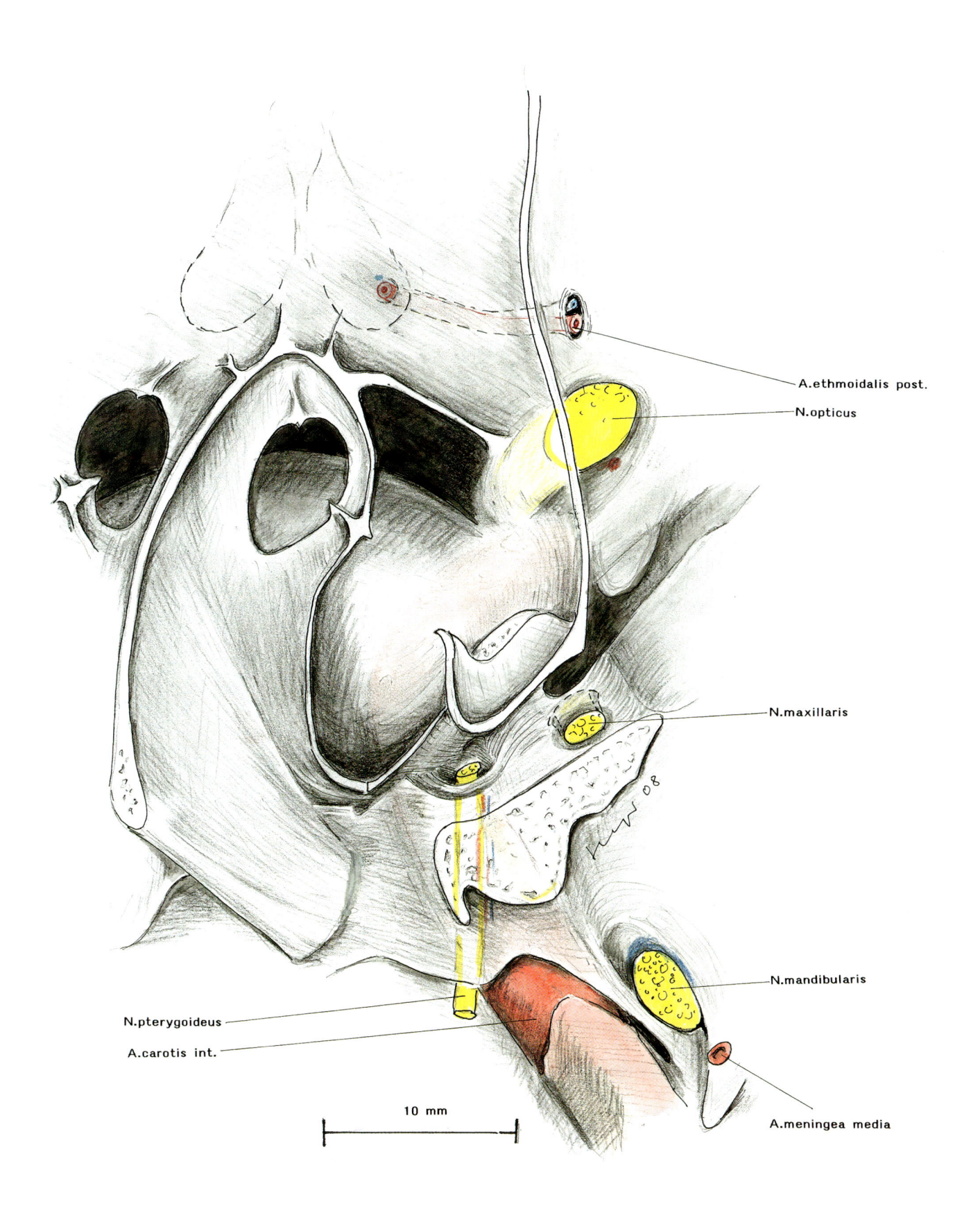

A.ethmoidalis post.

N.opticus

N.maxillaris

N.mandibularis

N.pterygoideus

A.carotis int.

A.meningea media

10 mm

Fig. 27

Sinus sphenoidalis. Possible width.

FIG. 27

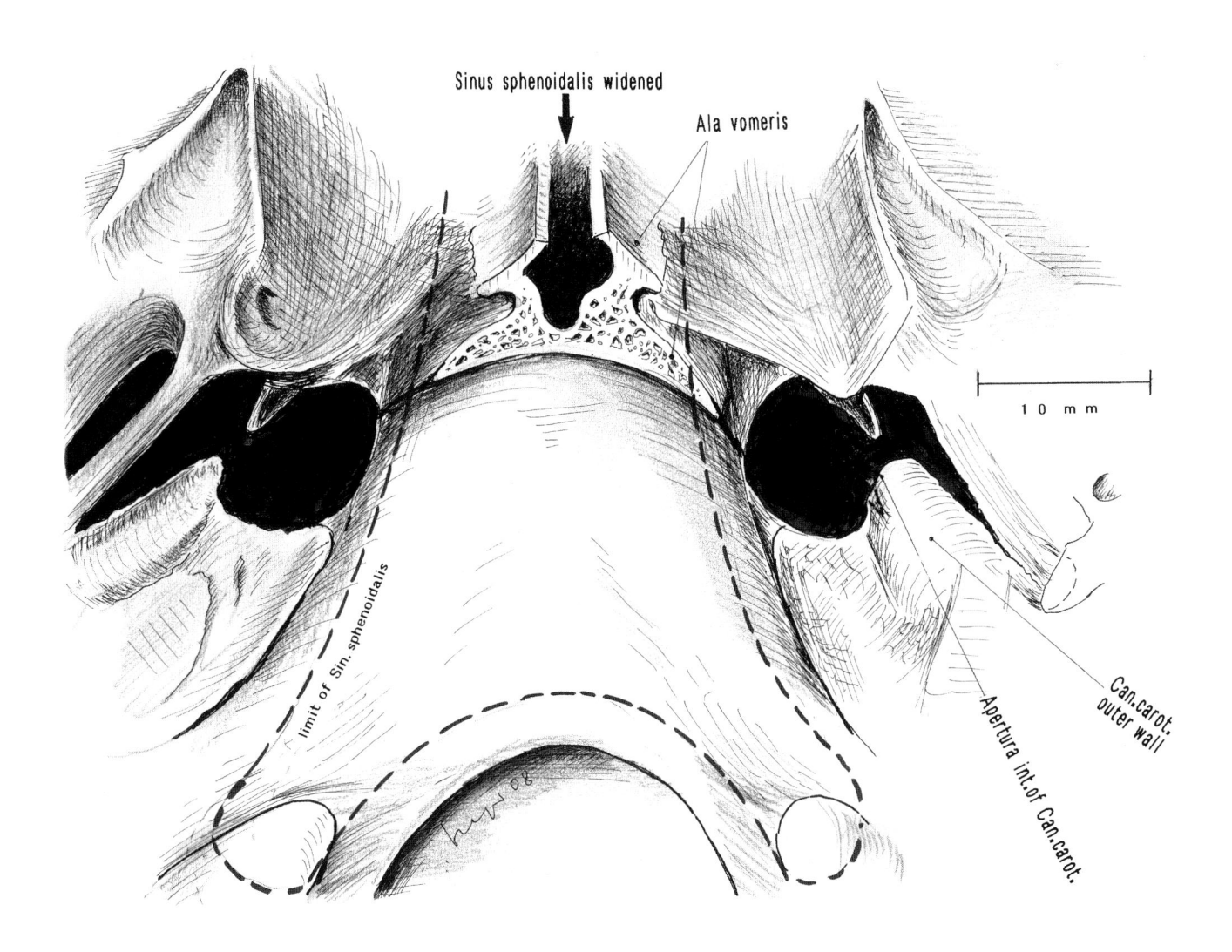

Sinus sphenoidalis widened

Ala vomeris

10 mm

limit of Sin. sphenoidalis

Can.carot.
outer wall

Apertura int.of Can.carot.

Fig. 28

View into a wide variant of Sinus sphenoidalis.

1.   Apertura interna of Canalis caroticus
2.   Foramen lacerum
3.   Sulcus caroticus

Arrow: Course of A. carotis int.

FIG. 28

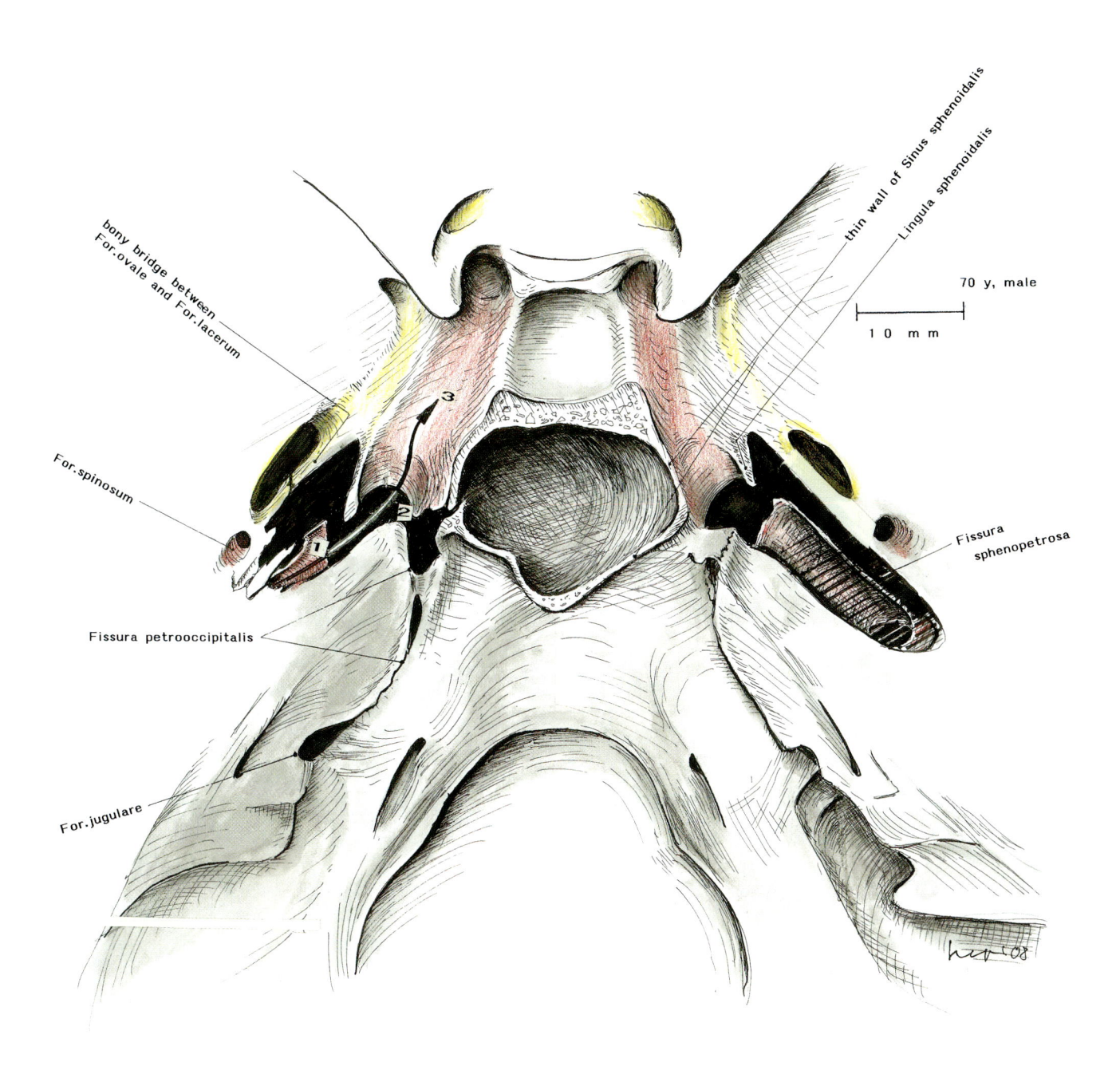

bony bridge between
For.ovale and For.lacerum

For.spinosum

Fissura petrooccipitalis

For.jugulare

thin wall of Sinus sphenoidalis

Lingula sphenoidalis

70 y, male

10 mm

Fissura
sphenopetrosa

Fig. 29

Sinus sphenoidalis

Topogram for Figs. 30 and 31

FIG. 29

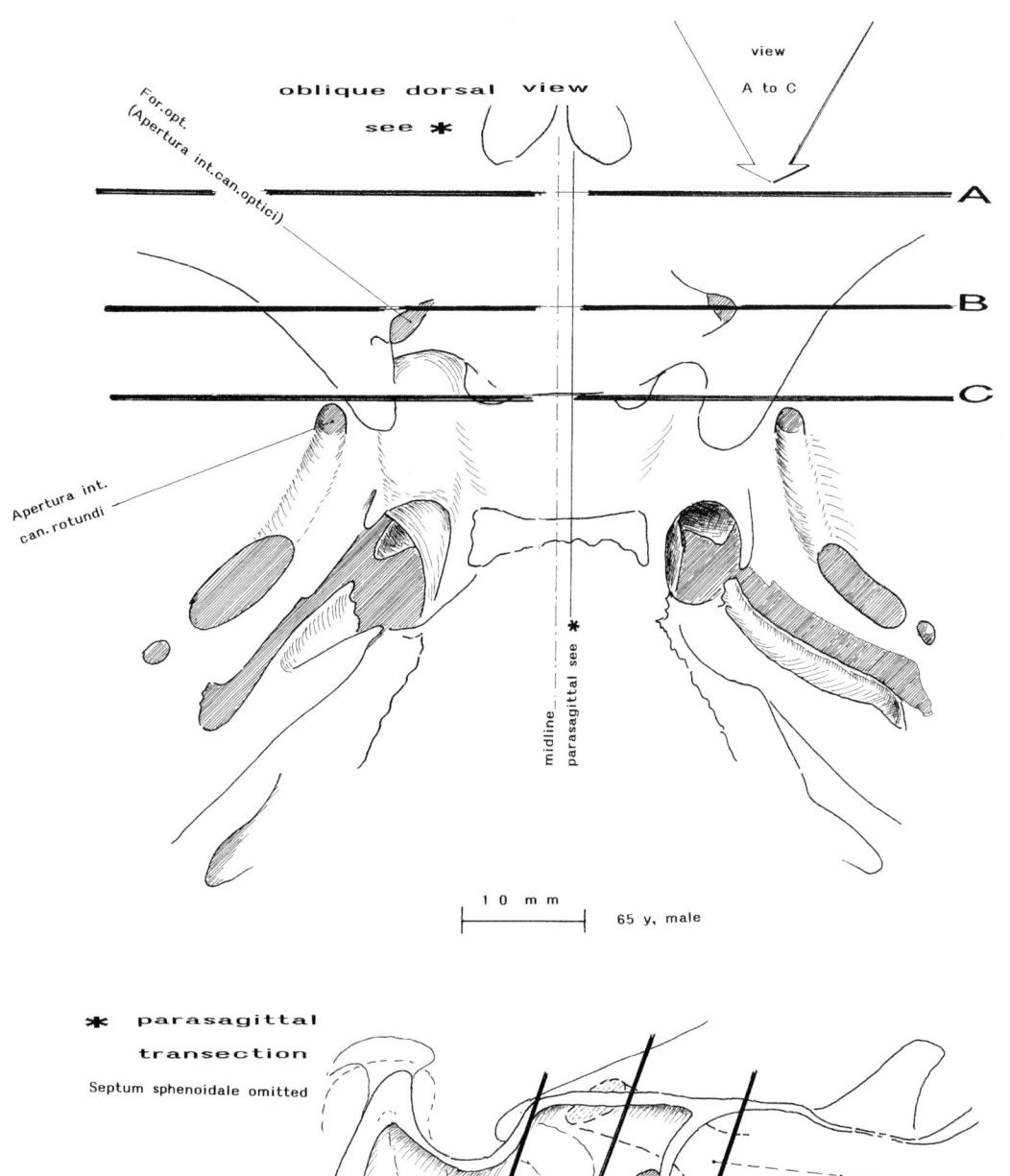

oblique dorsal view
see *

view
A to C

For.opt.
(Apertura int.can.optici)

A

B

C

Apertura int.
can.rotundi

midline
parasagittal see *

10 mm

65 y, male

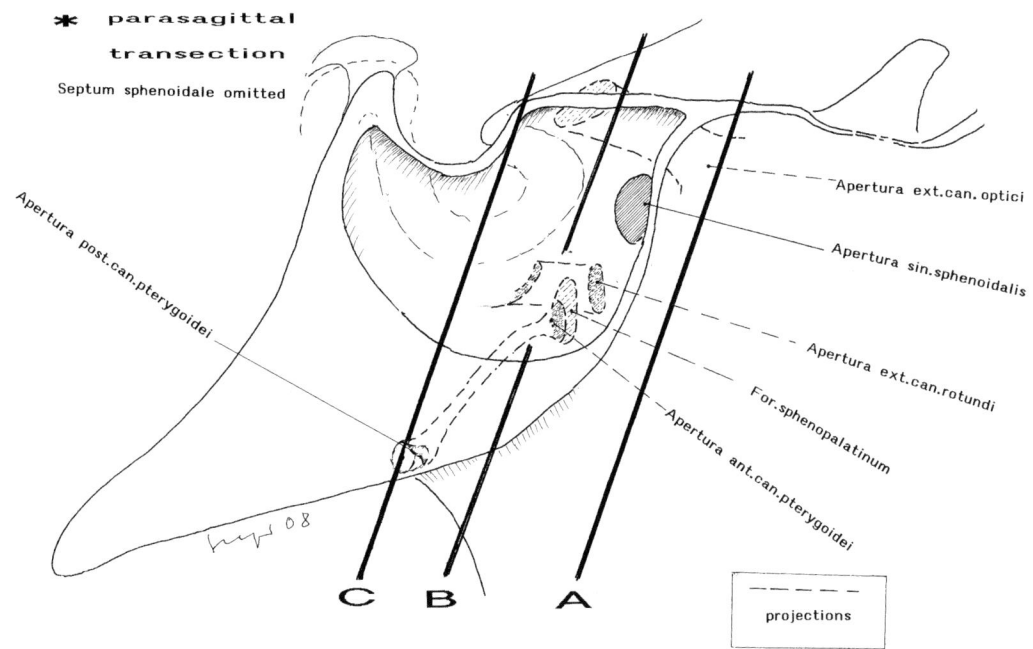

* parasagittal
transection
Septum sphenoidale omitted

Apertura post.can.pterygoidei

Apertura ext.can.optici

Apertura sin.sphenoidalis

Apertura ext.can.rotundi

For.sphenopalatinum

Apertura ant.can.pterygoidei

C B A

projections

Fig. 30

Sinus sphenoidalis

A   Vertical transection at the level of Recessus sphenoethmoidalis

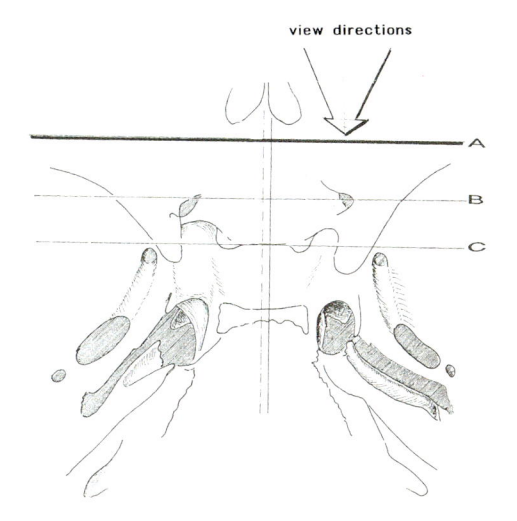

view directions

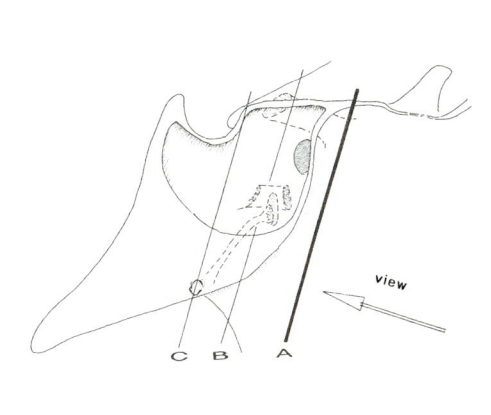

view

A

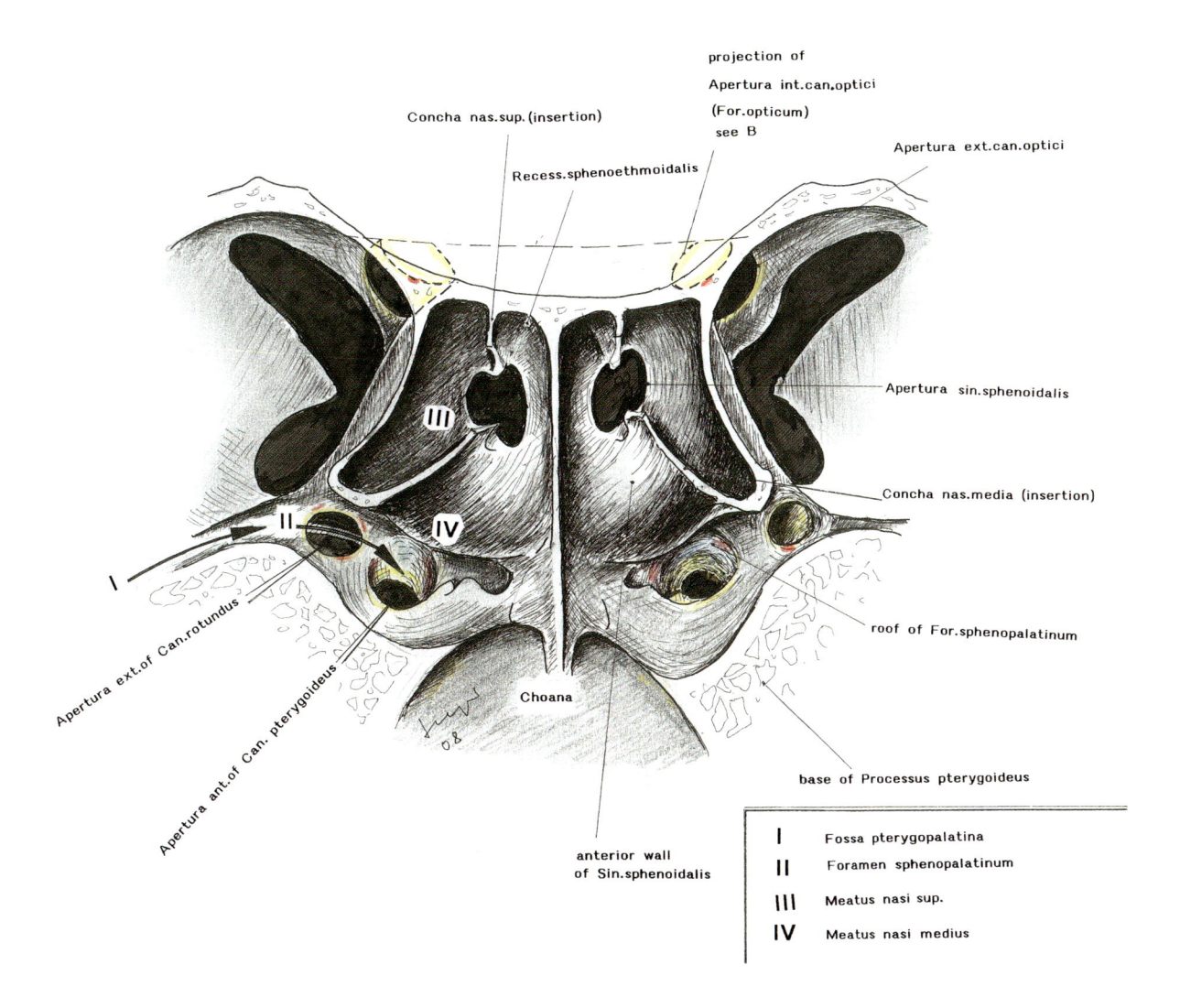

projection of
Apertura int.can.optici
(For.opticum)
see B

Concha nas.sup.(insertion)

Recess.sphenoethmoidalis

Apertura ext.can.optici

Apertura sin.sphenoidalis

Concha nas.media (insertion)

III

IV

II

I

roof of For.sphenopalatinum

Apertura ext.of Can.rotundus

Apertura ant.of Can. pterygoideus

Choana

anterior wall
of Sin.sphenoidalis

base of Processus pterygoideus

I	Fossa pterygopalatina
II	Foramen sphenopalatinum
III	Meatus nasi sup.
IV	Meatus nasi medius

Fig. 31

Continuation of Fig. 30

B Vertical transection at the level of For. opticum and posterior to Foramen sphenopalatinum

C Vertical transection at the level of Tuberculum sellae and at the anterior margin of Foramen lacerum.

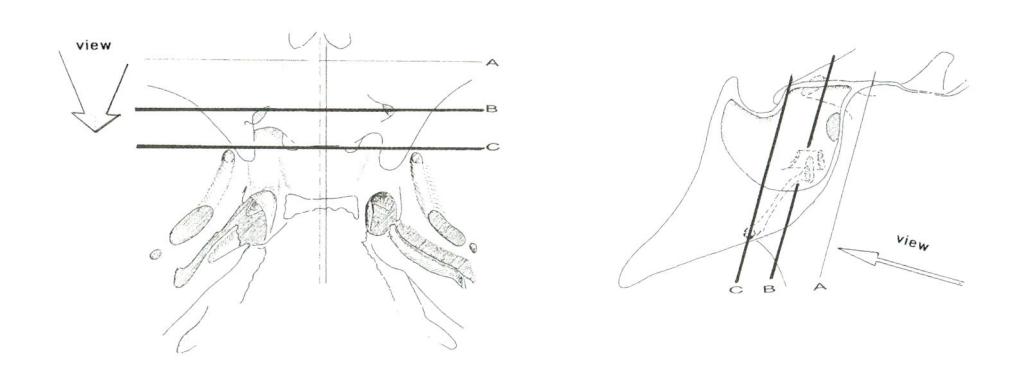

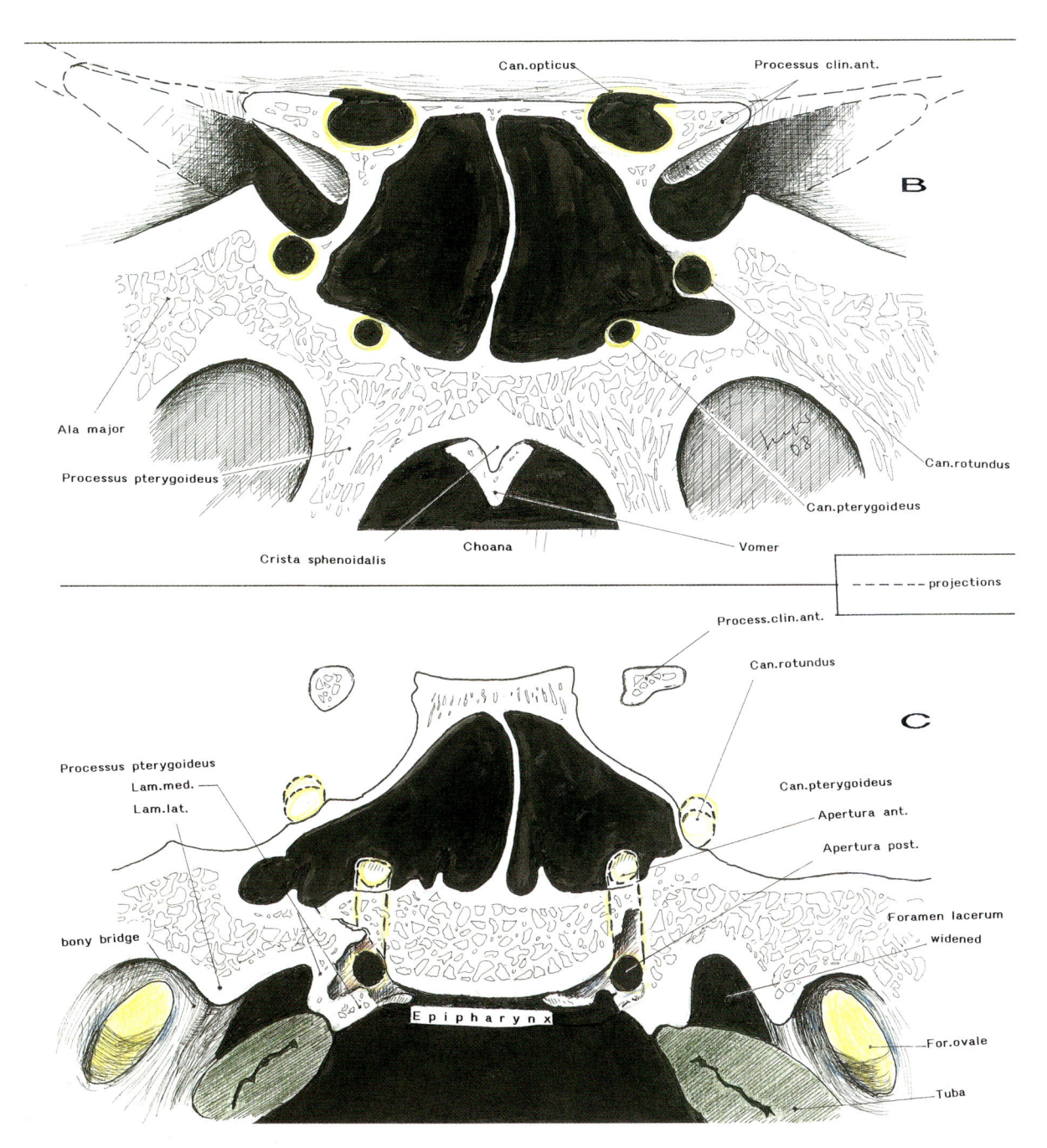

view

B

Can.opticus

Processus clin.ant.

Ala major

Processus pterygoideus

Crista sphenoidalis

Choana

Vomer

Can.pterygoideus

Can.rotundus

- - - - - projections

C

Process.clin.ant.

Can.rotundus

Processus pterygoideus
Lam.med.
Lam.lat.

Can.pterygoideus

Apertura ant.

Apertura post.

bony bridge

Foramen lacerum
widened

Epipharynx

For.ovale

Tuba

Fig. 32

Sinus sphenoidalis. Its roof is removed. Os basilare is transected.

Abbreviations
1 medial basal segment of Fissura orbitalis (located immediately lateral from the knee of Curvatura ant. of A. carotis int.)
2 Processus clinoideus ant.
3 Canalis opticus
4 Concha nasalis sup., beyond Apertura sphenoidalis
5 Meatus nasi superior beyond Apertura sphenoidalis
6 as 3
7 Foramen ovale
8 fenestrated wall of Sinus sphenoidalis
9 bony ridges of the floor of Sinus sphenoidalis
10 as 9
11 Apex of the petrous bone
12 Apertura int. of Foramen rotundum

FIG. 32

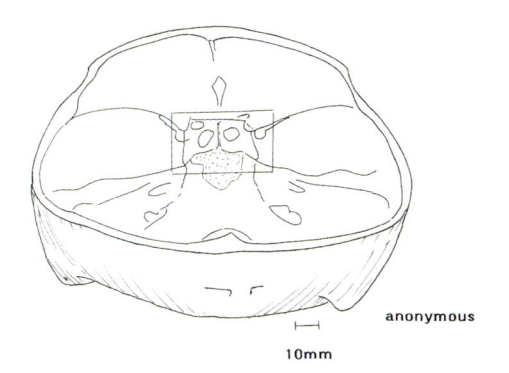

anonymous

10mm

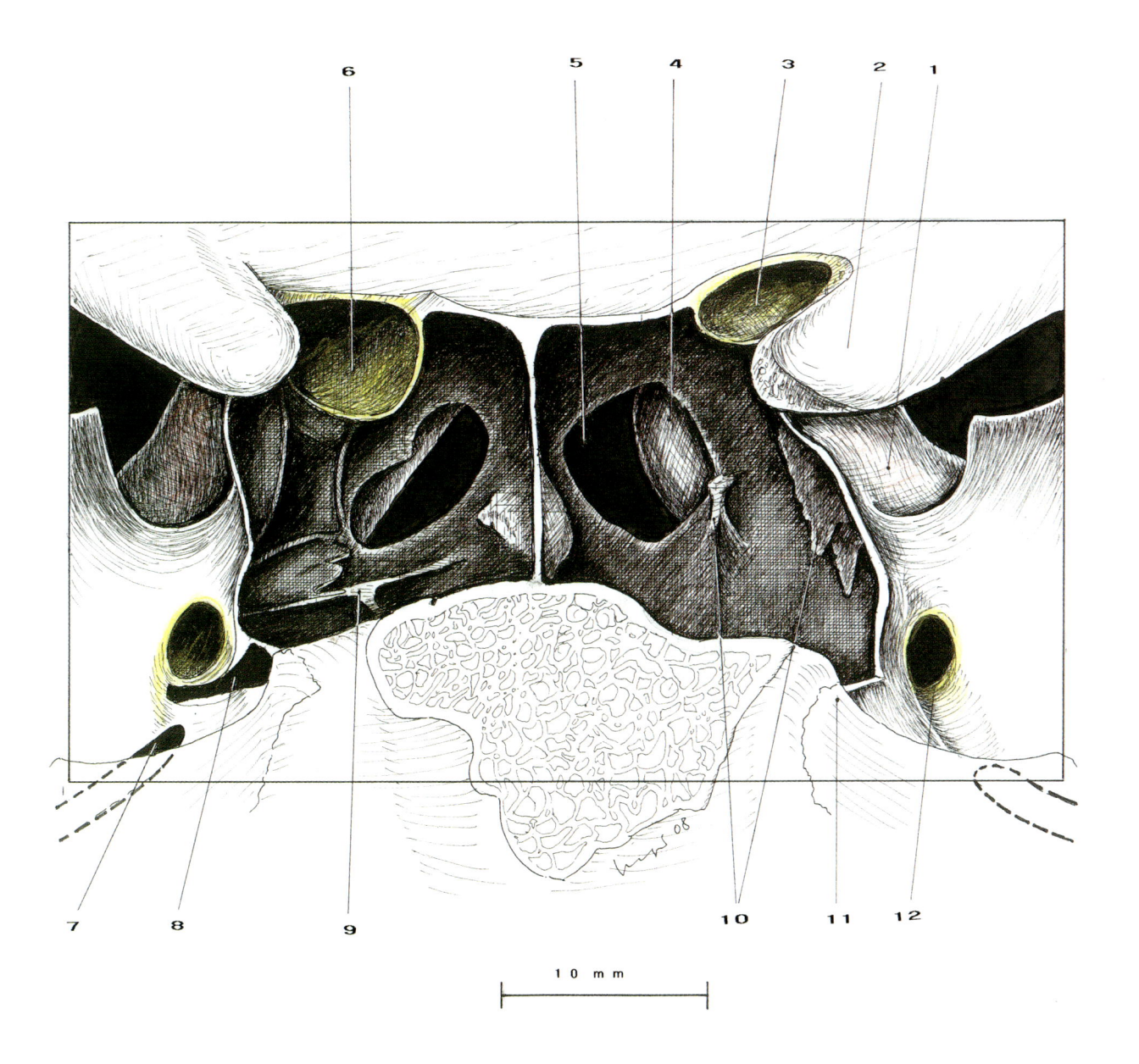

10 mm

Fig. 33

Continuation of Fig. 32
A leftsided pneumatization is depicted in this cadaver skull dissection. This type of variant may be connected to a widened Sinus sphenoidalis or to a widened posterior ethmoid cell (Lang 1981, p 457). These connections may present as extensions of a pneumatized roof of Processus clinoideus anterior or as a pneumatization and bony duplication of the roof of Canalis opticus, or both. This is demonstrated in Fig. 20.

Abbreviations
1 Apex of the petrous bone
2 Foramen ovale
3 Lamina medialis of Processus pterygoideus beyond Foramen lacerum
4 bony sulcus between Canalis rotundus and Foramen ovale, variable 4a as 4, fenestrated
5 Foramen opticum
6 Sulcus chiasmatis
7 Apertura sinus sphenoidalis
8 Canalis opt.
9 Septum of Sinus sphenoidalis

FIG. 33

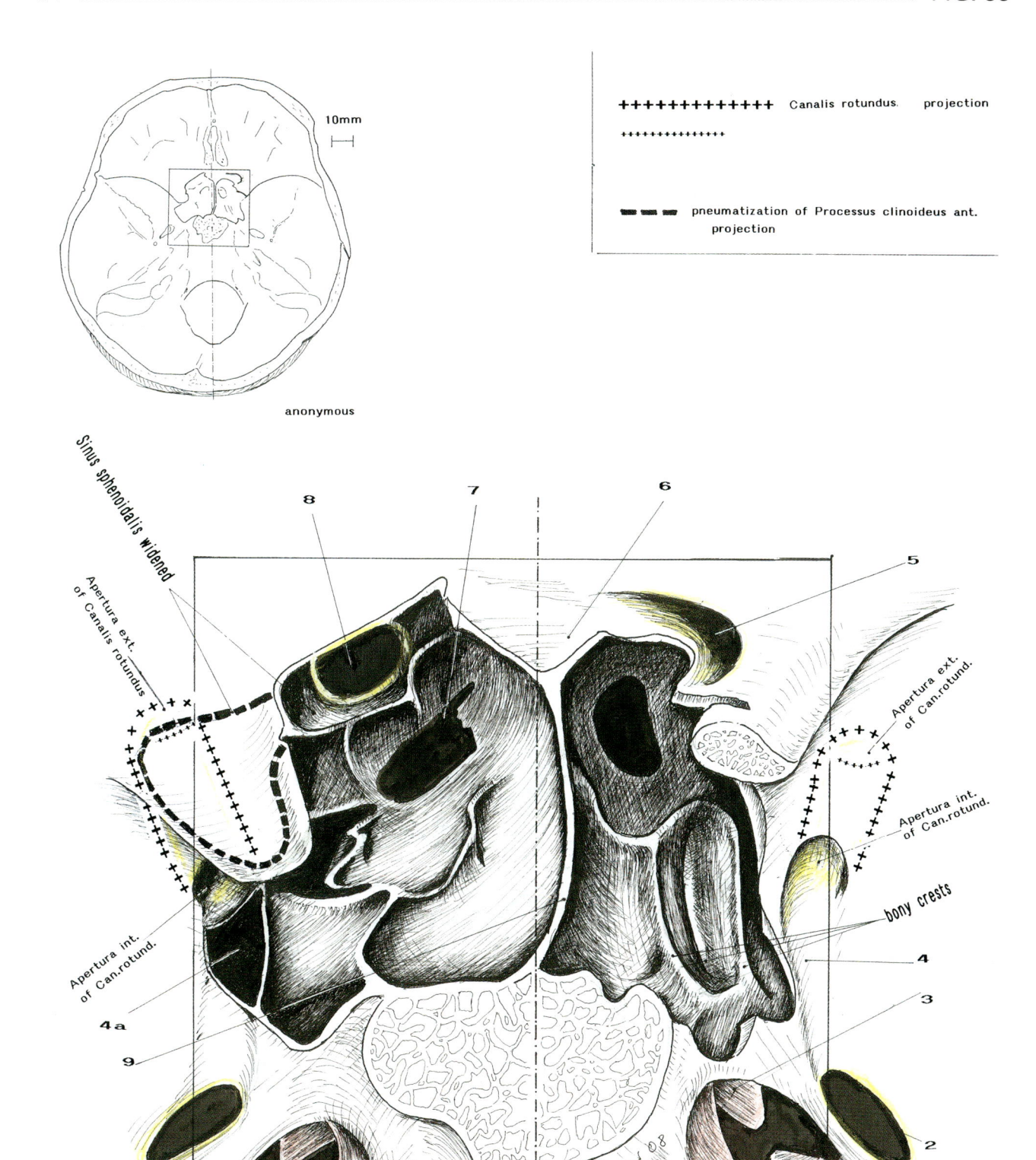

++++++++++++++   Canalis rotundus.   projection

+++++++++++++++

▬ ▬ ▬ ▬ ▬   pneumatization of Processus clinoideus ant.
projection

anonymous

10mm

Sinus sphenoidalis widened

Apertura ext.
of Canalis rotundus

8

7

6

5

Apertura ext.
of Can.rotund.

Apertura int.
of Can.rotund.

bony crests

4

3

Apertura int.
of Can.rotund.

4a

9

2

1

10 mm

Fig. 34

Axial transection of Cavum nasi, Sinus sphenoidalis and surrounding structures. Schematic presentation.

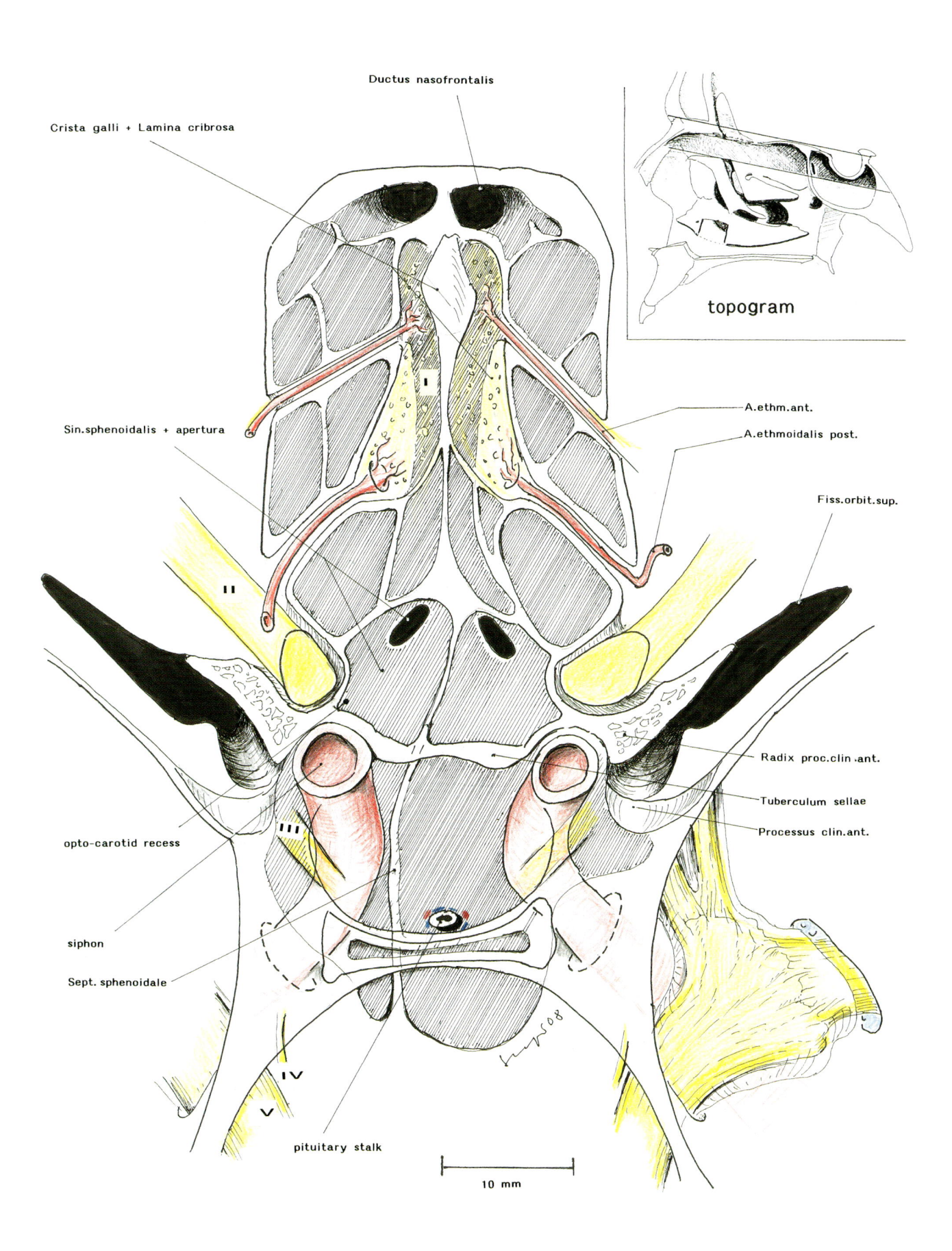

Ductus nasofrontalis

Crista galli + Lamina cribrosa

Sin.sphenoidalis + apertura

A.ethm.ant.

A.ethmoidalis post.

Fiss.orbit.sup.

Radix proc.clin.ant.

Tuberculum sellae

Processus clin.ant.

opto-carotid recess

siphon

Sept. sphenoidale

pituitary stalk

topogram

10 mm

Literature

Boudin G, Laude M, Freche Ch (1982) Anatomical bases for the surgical approach to the Vidian nerve (in Canalis pterygoideus) Anat Clin 3, 239–242

Castelnuovo P, Delue G, Locatelli D, Padoan G, de Bernardi F, Pistochini A, Bignami M (2006) Endonasal endoscopic duraplasty. Our experience. Skull base 16 (1), 15–8

Castelnuovo P, Locatelli D, Mauri S, de Bernardi F (2003) Extended endoscopic approaches to the skull base, anterior cranial base CSF leaks. In: De Divitis E, Cappabianca P (ed): Endoscopic endonasal transsphenoidal surgery. Springer, Wien New York, pp 137–138

Castelnuovo P, Locatelli D, Santi L, Emanuelli E, Pagella F, Canevari FR (1998) Sinonasal endoscopic access to the pituitary gland. In: Stamberger, Wolf eds. ERS & ISIAN Meeting. Monduzzi, pp 337–339

Cavallo LM, Messina A, Cappabianca P, Esposito F, de Divitis E, Gardner P, Tschabitscher M (2005). Endoscopic endonasal surgery of the midline skull base. Anatomical study and clinical considerations. Neurosurg Focus 19 (1), E2, pp 1–14

De Divitis E, Cappabianca P, Cavallo LM (2002) Endoscopic transethmoidal approach: adaptility of the procedure to different sellar lesions. Neurosurgery 51 (3), pp 699–705

Fujji K, Chambers SM, Rhoton AL (1979) Neurovascular relationship of the sphenoid sinus. A microsurgical study. J Neurosurg 50, 31–39

Gibo H, Lenkey C, Rhoton AL (1981) Microsurgical anatomy of the supraclinoid portion of the carotid artery. J Neurosurg 55, 560–574

Grisoli F, Vincentelli F, Henry J (1982) Anatomical bases for the transsphenoidal approach to the pituitary gland. Anat Chir 3, 207–220

Kassam AB, Gardner P, Snyderman C, Mintz A, Carrau R (2005) Expanded endonasal approach: Fully endoscopic, completely transnasal approach to the middle third of the clivus, petrous bone, middle cranial fossa, and infratemporal fossa. Neurosurg Focus Jul; 15:19(1):E6

Krmpotic-Nemanic J, Draf W, Helms J (1985) Chirurgische Anatomie des Kopf-Hals-Bereichs p 231, p 233. Springer, Berlin Heidelberg New York Tokyo

Lang J (1981) Neuroanatomie der Nn. opticus, trigeminus, facialis, glossopharyngeus, vagus, accessorius und hypoglossus. Arch Otorhinolaryngol 231, 1–69

Lang J, Schlehan FA (1981) Über die postnatale Entwicklung der Fissurae orbitales. Gegenbaurs Morphol Jahrb 127, 849–859

Locatelli D, Rampa F, Acchiardi I, Bignami M, de Bernardi F, Castelnuovo P (2006)

Endoscopic endonasal approaches for repair of cerebrospinal fluid leaks. Nine-year experience. Neurosurgery 58 (Suppl 2). ONS 246–256

Mandiola E, Delgado A, Chatain I (1979). Contribucion al studio de la fossa pterigopalatina/I Contenido vascular y tecnica de abordaje por la via transantral. Rev Otorhinolaryngol 39, 39–46

Nicolic V (1967). Variations du trou sphenopalatin. Acta anat 68. 189–198

Osborn AG (1980) The vidian artery: Normal and pathological anatomy. Radiology 136, 373–378

Rauber-Kopfsch (1906) Bd. 2, p 245. Thieme, Leipzig

Rauber-Kopsch (1908) Bd. 6, p 1017. Thieme, Leipzig

Rose KG, Ortmann R, Seegers D (1979) Vidian neurectomy: neuroanatomical considerations and a report on a new surgical approach. Arch Otorhinolaryngol 224, 157–168

Turvey TA, Fonseca RJ, Hill CH (1980) The anatomy of the intermaxillary artery in the pterygopalatine fossa: Its relationship to maxillary surgery. J Oral Surg 38, 92–95

CHAPTER IV
TUBA AUDITIVA (EUSTACHII)
(Figs. 35 to 40)

Overview (Figs. 35 und 36)

Tuba auditiva is divided into Pars cartilaginea and Pars ossea.
Pars cartilaginea is located extracranially, Pars ossea is enclosed by Pyramis

The long course of Tuba extends from its Orificium at Pharynx to its Ostium at Cavum tympani. The entrance of Pars cartilaginea at Pyramis is called Apertura tubae. Here Tuba is narrowed (Isthmus tubae). Between this point and Cavum tympani, the thin-walled fine Semicanalis musculotubarius encloses M. tensor tympani (upper segment) and Tuba (inferior segment).

Pars cartilaginea

Orificium tubae is located in the axial level of Concha inferior. The posterior wall of Tuba, Torus tubarius, is located between Orificium and Recessus pharyngeus (Rosenmuelleri). The bony bed of Tuba begins caudally at Fossa scaphoidea. This is a small triangular fovea at the dorsal base of Lamina medialis of Processus pterygoideus, anterior to Foramen lacerum. At Foramen lacerum, Tuba and A.carotis int. are separated by cartilagineous and fibrous layers. Posterior to Foramen lacerum the Tuba is running along its extracranial bed, Fissura sphenopetrosa, parallel to the intrapetrosal (pyramidal) course of A. carotis int. (Fig. 37). Fissura sphenopetrosa often presents as a wide gap, which forms an extension of Foramen lacerum into a lateral-posterior direction. It may extend to the Spina angularis-Apertura-tubae-complex. The gap of Fissura sphenopetrosa may connect to the longitudinal dorsal gap of Canalis caroticus. Tuba and A. carotis int. are interposed between the extra- and intracranial surface of the cranial base. Due to this proximity inexact punctures of Foramen ovale may be dangerous! Foramen ovale is located close to Tuba. If Tuba is punctioned (instead of Foramen ovale), injury to A. carotis interna is possible with consecutive hematotympanon. This was a well known complication in the past, in electrocoagulation of Ganglion Gasseri (Schenk and Seeger, 1968). Foramen spinosum is located lateral to Tuba. The extracranial segment of A. meningea media and Mm.tensor and levator veli palatinae, which originates from Tuba and from Spina angularis are in close proximity. They may mask Apertura tubae (Fig. 50, see chapter 5).

Pars ossea (C in Fig. 37, and Fig. 40)

Apertura tubae of Pyramis is covered by Spina angularis of Ala major and palatinal muscles, close to Foramen spinosum and its content, A. meningea media. The distance between to the Apertura externa of the carotid channel is variable.
Tuba crosses the bending segment of the carotid channel at its dorsal point, close to Apertura tubae.

TUBA AUDITIVA (EUSTACHII) (Figs. 35 to 40)

Fig. 35

Overview

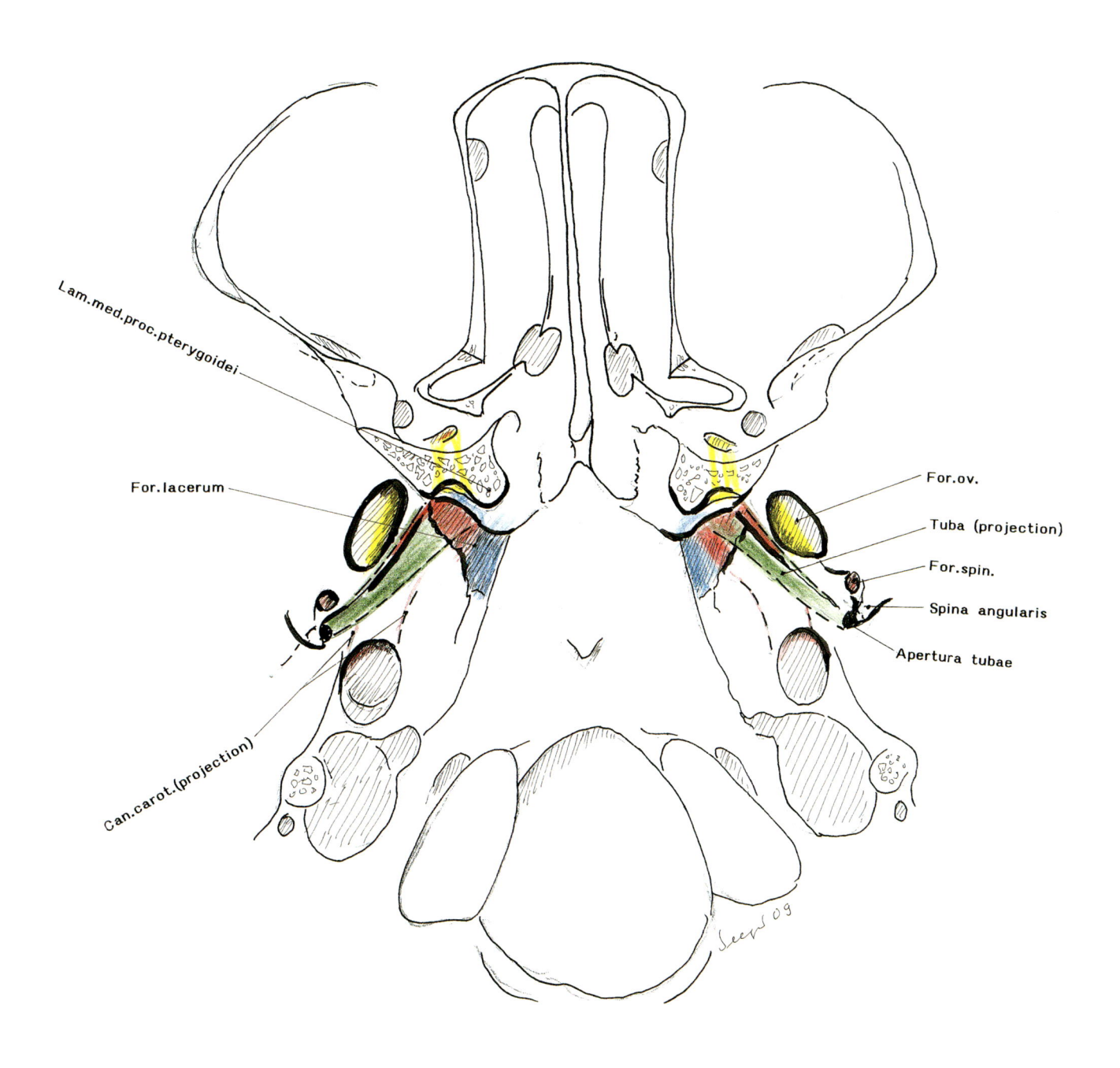

Fig. 36

Tuba auditiva (Eustachii), overview

A Schematic illustration
B Basal view according to Spalteholz (1907). Details added.
Pars cartilaginea tubae split and everted (arrows)

Abbreviations
1 Tunica mucosa tubae auditivae (Eustachii)
2 Pars membranacea
3 Pars cartilaginea
4 M. levator veli palatini
5 M. tensor veli palatini
6 M. tensor tympani
7 Meatus acusticus ext., Pars ossea

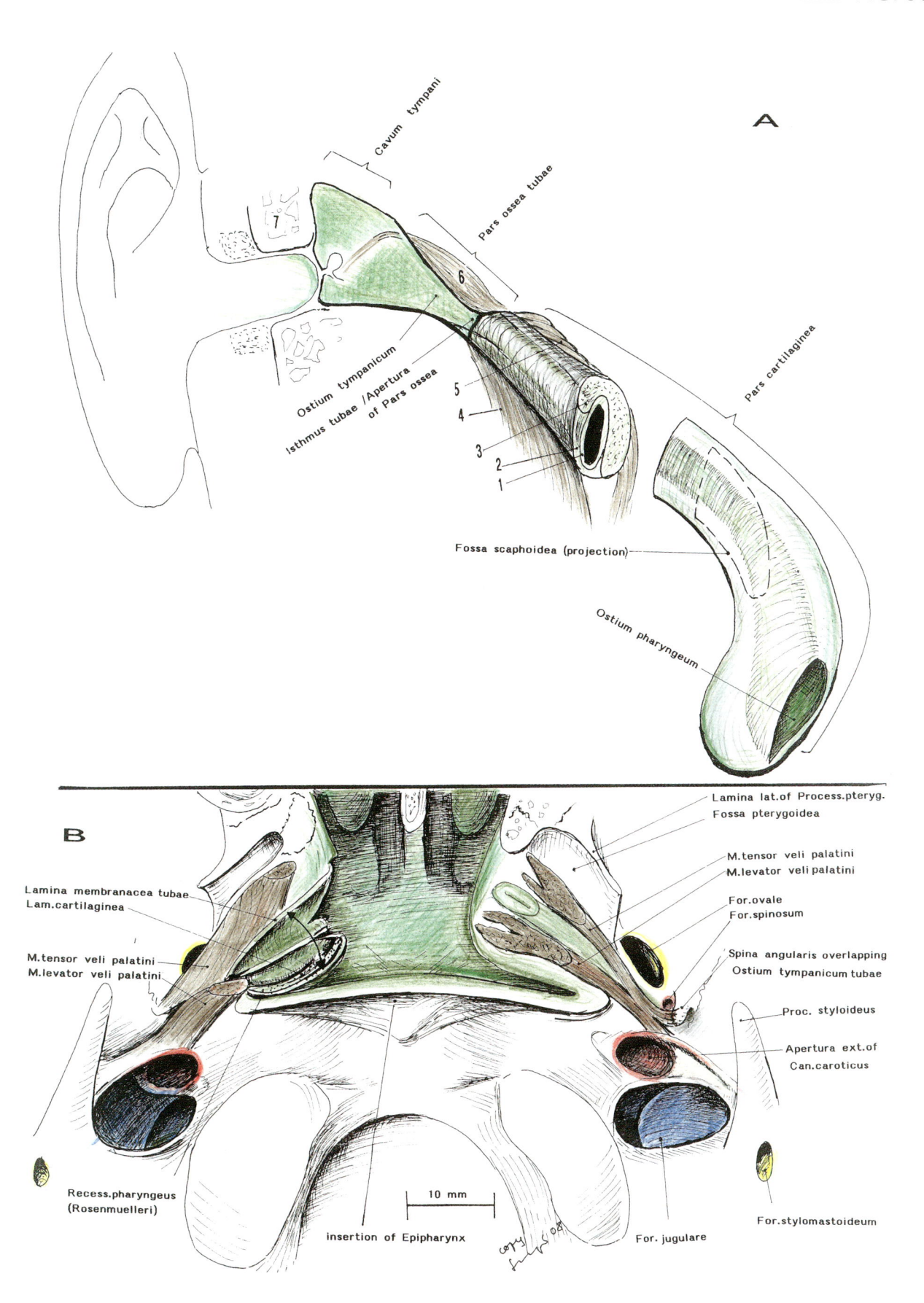

A

Cavum tympani

Pars ossea tubae

Ostium tympanicum

Isthmus tubae / Apertura of Pars ossea

Pars cartilaginea

5
4
3
2
1

Fossa scaphoidea (projection)

Ostium pharyngeum

B

Lamina lat.of Process.pteryg.
Fossa pterygoidea

M.tensor veli palatini
M.levator veli palatini

For.ovale
For.spinosum

Spina angularis overlapping
Ostium tympanicum tubae

Proc. styloideus

Apertura ext.of
Can.caroticus

For.stylomastoideum

Lamina membranacea tubae
Lam.cartilaginea

M.tensor veli palatini
M.levator veli palatini

Recess.pharyngeus
(Rosenmuelleri)

insertion of Epipharynx

For. jugulare

10 mm

Fig. 37

Tuba auditiva (Eustachii), details

A Transectional planes A1 to A4
B Locations of A1 to A4
C Apertura tubae

FIG. 37

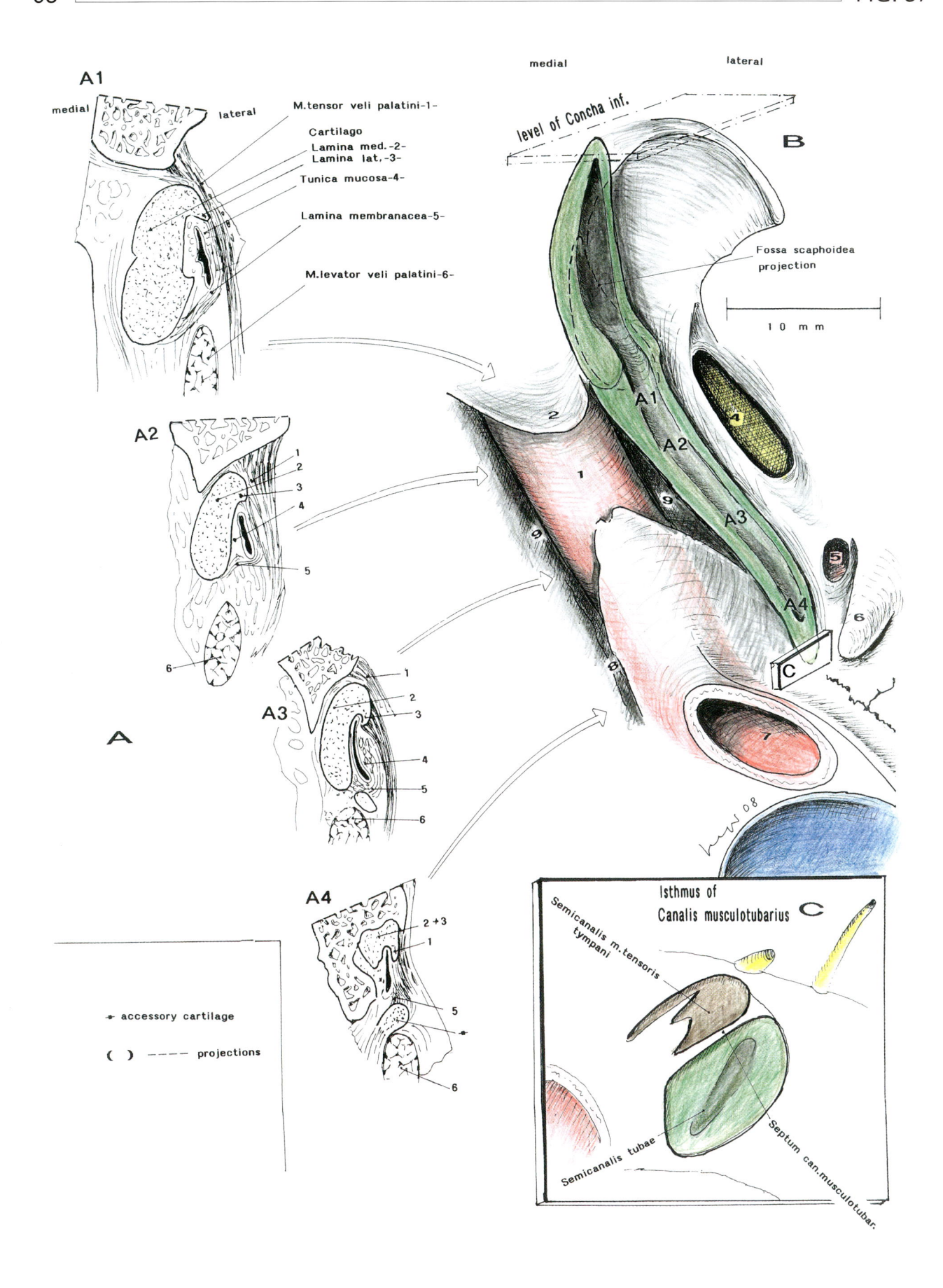

A1

medial   lateral

M.tensor veli palatini-1-

Cartilago
Lamina med.-2-
Lamina lat.-3-

Tunica mucosa-4-

Lamina membranacea-5-

M.levator veli palatini-6-

A2

1
2
3
4

5

6

A3

1
2
3

4
5
6

A

A4

2 + 3
1

5

6

→ accessory cartilage

( ) ----- projections

medial   lateral

level of Concha inf.

B

Fossa scaphoidea
projection

10 mm

A1
A2
A3
A4

2
1
9
9
8
5
6
7
08

C

Isthmus of
Canalis musculotubarius   C

Semicanalis m.tensoris
tympani

Semicanalis tubae

Septum can.musculotubar.

Fig. 38

Tuba, topographical relationship to Epipharynx and to anterior segment (orificium) of Tuba at the level of Concha inf. According to Spalteholz (1906, p 504), modified.

Landmarks
1   Vomer
2   Processus vaginalis (see Chapter 5)
3   Os petrosum
4   Sulcus petrosus inf.
5   Clivus
6   Pars basilaris

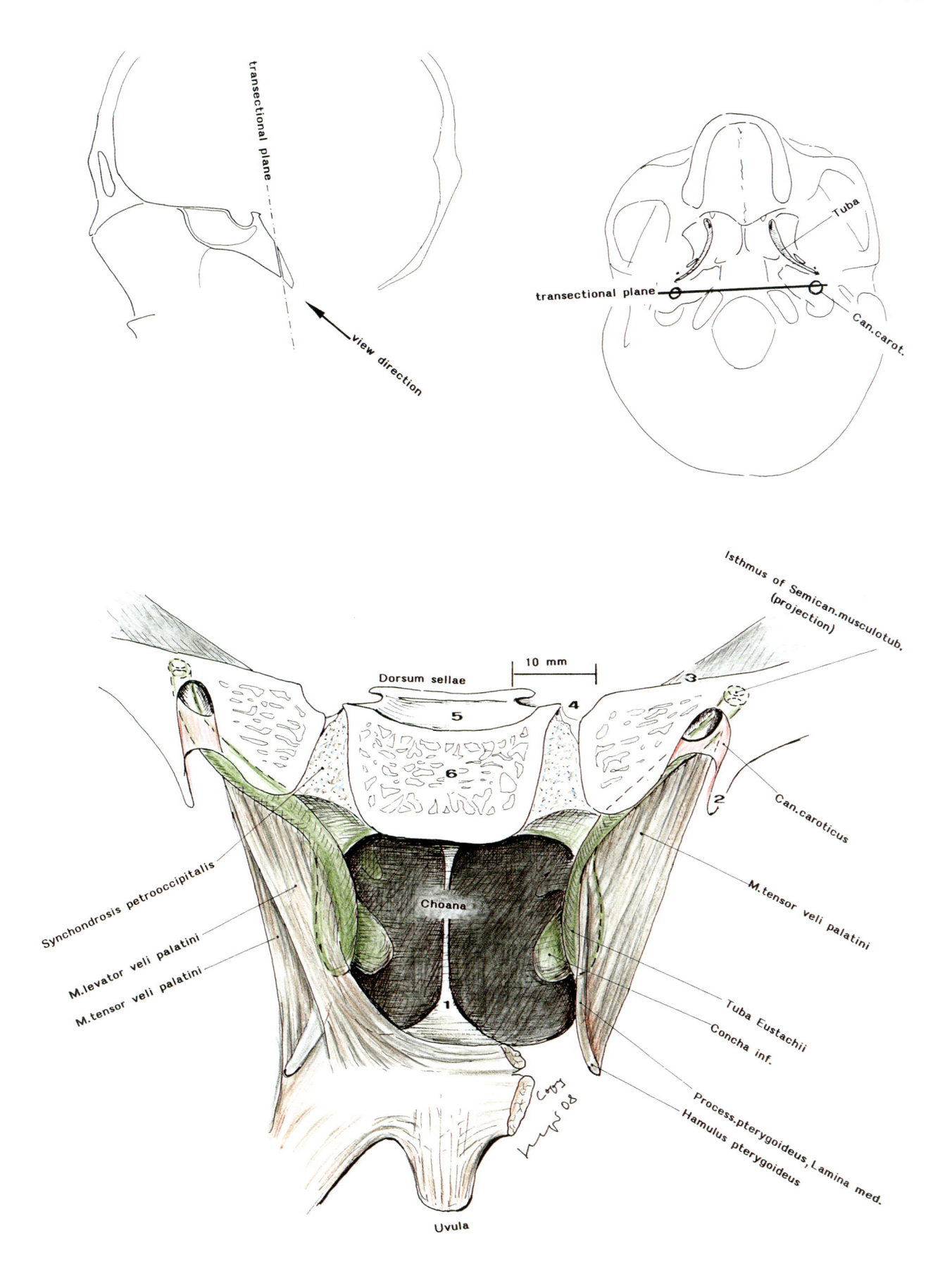

transectional plane

view direction

Tuba

transectional plane

Can.carot.

Isthmus of Semican.musculotub.
(projection)

10 mm

Dorsum sellae

5

4

3

6

2

Choana

Can.caroticus

Synchondrosis petrooccipitalis

M.tensor veli palatini

M.levator veli palatini

M.tensor veli palatini

1

Tuba Eustachii

Concha inf.

Process.pterygoideus, Lamina med.

Hamulus pterygoideus

Uvula

Fig. 39

Continuation of Fig. 38

A Bed of Tuba
B Proximal segment of Tuba
C Overview

Abbreviations
1 Condylus occipitalis
2 Processus styloideus
3 Vagina of Processus styloideus
4 Porus acusticus ext.
5 Fossa mandibularis and Fissura petrotympanica Glaseri
6 Tuberculum articulare of Os temporale
7 Fossa pterygopalatina
8 spine between Facies temporalis and infratemporalis of Ala major of Os sphenoidale
9 Fossa pterygoidea

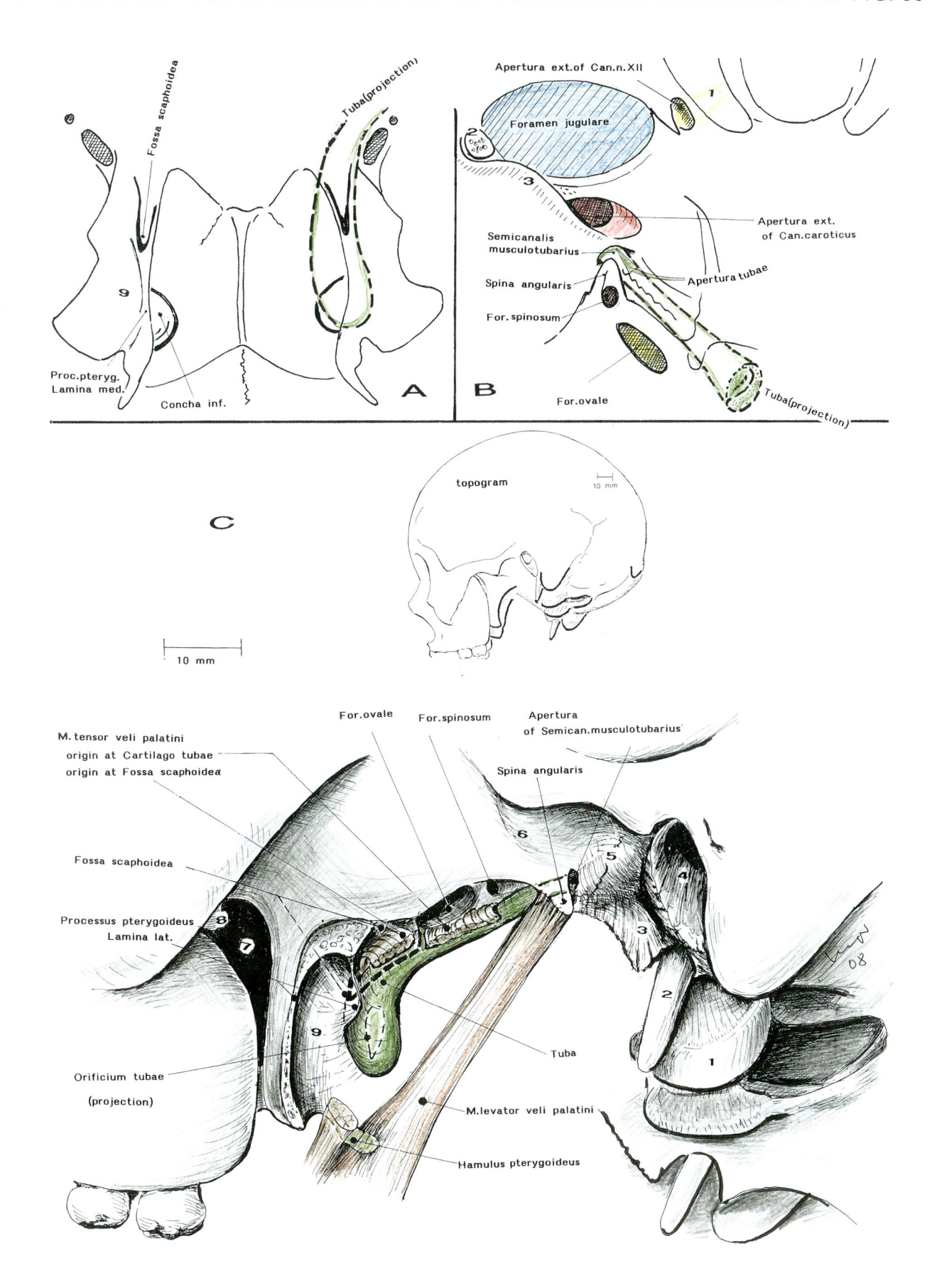

A

Fossa scaphoidea

Tuba(projection)

9

Proc.pteryg.
Lamina med.

Concha inf.

B

Apertura ext.of Can.n.XII

Foramen jugulare

Semicanalis
musculotubarius

Spina angularis

For. spinosum

For.ovale

Apertura ext.
of Can.caroticus

Apertura tubae

Tuba(projection)

C

topogram

10 mm

10 mm

M. tensor veli palatini
origin at Cartilago tubae
origin at Fossa scaphoidea

For.ovale   For.spinosum   Apertura
of Semican.musculotubarius

Spina angularis

Fossa scaphoidea

Processus pterygoideus
Lamina lat.

Orificium tubae

(projection)

Tuba

M.levator veli palatini

Hamulus pterygoideus

Fig. 40
Tuba auditiva, Pars ossea.
Historical cadaver head and skull dissections, according to Spalteholz (1907, 803ff, 810), slightly modified

A Horizontal transection
B Vertical coronal and vertical oblique transections combined
C Coronal transection
D Sagittal presentation of Cavum tympani, medial wall

Abbreviations
1. Pars cartilaginea of Meatus acusticus ext.
2. Meatus acusticus ext., Pars ossea
3. temporobasal cerebral cortex
4. A. carotis int.
5. Ostium tympanicum tubae
6. Cavum tympani
7. Membrana tympani
8. Sinus sigmoideus
9. M. levator veli palatinae
10. Tuba Eustachii, Pars membranacea
11. Tuba Eustachii, Pars cartilaginea
12. Ostium pharyngeum tubae
13. Isthmus tubae
14. Processus mastoideus
15. Processus styloideus
16. Processus vaginalis
17. Cochlea
18. Fundus of Meatus acusticus int.
19. Recessus epitympanicus
20. as 5, Semicanalis tubarius
21. Semicanalis muscularis
22. M. tensor tympani
23. Promontorium
24. Antrum tympanicum
25. Outer wall of Canalis n. facialis,
Between 24 and 35: Recessus epitympanicus

FIG. 40

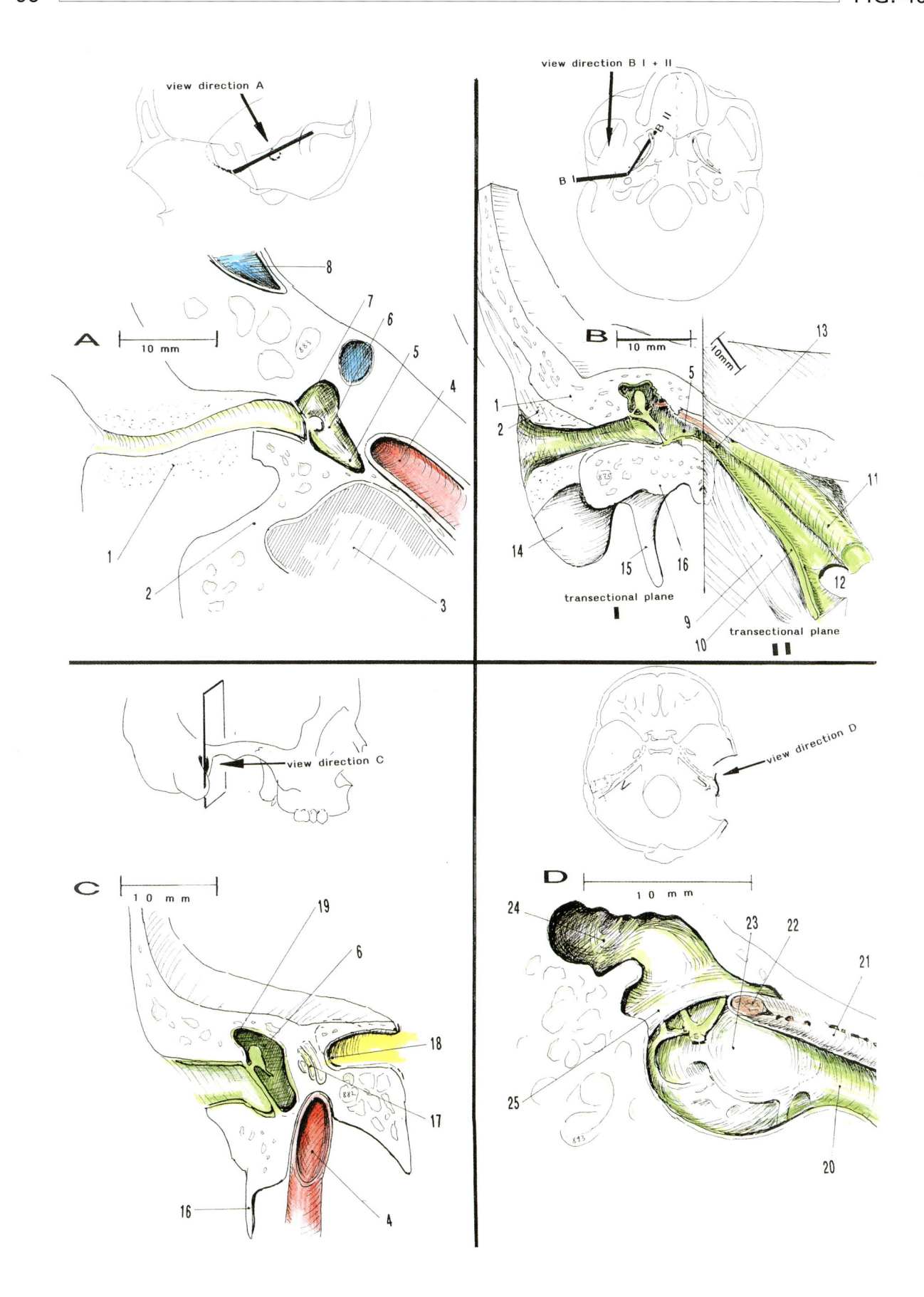

view direction A

view direction B I + II

B II

B I

A 10 mm

8

7

6

5

4

1

2

3

B 10 mm

10mm

13

1

2

5

14

15

16

9

10

11

12

transectional plane

transectional plane

view direction C

view direction D

C 10 mm

19

6

18

17

16

4

D 10 mm

24

23

22

21

25

20

Literature

Krmpotic-Nemanic J, Draf W, Helms J (1985) Chirurgische Anatomie des Hals-Kopf-Bereichs. P 207. Springer, Berlin Heidelberg New York Tokyo

Schenk D, Seeger W (1968) Otologische Komplikationen nach Kirschnerscher Elektrokoagulation des Ganglion Gasseri. Tagung Dtsch. Ges. Hals-Nasen-Ohrenheilk. Bad Reichenhall, 27.5.

Spalteholz W (1906) Handatlas der Anatomie des Menschen, Bd. 3, pp 502-509, p 803, p 810, p 819, p 822. Hirzel, Leipzig

Spalteholz W (1907) Handatlas der Anatomie des Menschen, Bd. 1, p 12. Hirzel, Leipzig

Thomas JR (1980) Tympanic neurectomy and chorda tympani section. Aust NZJ Surg 50, 352–355

CHAPTER V
PYRAMIS (PETROUS BONE,
PARS PETROSA PLUS PARS TYMPANICA)
(Figs. 41 to 63)

Overview (Figs. 41 to 43)

The exact outline of Pars petrosa and Pars tympanica is given by sutures in childhood
The pyramis resembles an irregular shaped pyramid with 3 planes and edges.
Facies posterior and anterior pyramidis are located on the inside of the skull base.
Facies inferior is located on the outside of the skull base.

Pyramis, schematic drawing (Figs. 42 and 43)

Facies posterior pyramidis (Fig. 42)

Central structures are Porus and Meatus acusticus internus. Margo sup. pyramidis, Fissura petrooccipitalis and Apex pyramidis forms the logs of pyramis. The base of Pyramis is not exactly defined (Fig. 42). Margo sup. pyramidis extends from Foramen lacerum to Sulcus sigmoideus (close to the sinus knee). Fissura petrooccipitalis extends from Foramen lacerum to the anterior margin of Foramen jugulare. At the base of Foramen jugulare, a fine suture separates Pyramis from Os occipitale. Foramen jugulare consists of a medial portion, which is a segment of the occipital bone, and a lateral portion, which is a segment of Pyramis.
The lateral margin of Foramen jugulare is divided into a small anterior and a wider posterior segment. They are divided by a small protrusion, Processus intrajugularis. This is not yet recognized as an exact neuronavigatory landmark.

Facies anterior pyramidis (Fig. 43)

The main structures are located on the margins. These are the horizontal segment of Canalis caroticus, Apertura tubae and adjacent structures of the sphenoid bone. These structures are located close to Fissura sphenopetrosa, between Ala major and Pyramis. Eminentia arcuata encloses Ductus semicircularis superior.

Facies inferior pyramidis (Fig. 43)

Its main structures are Apertura ext. of the carotid channel, Foramen jugulare, and Foramen stylomastoideum. Fissura sphenopetrosa and Fissura petrooccipitalis form the bolders. Processus vaginalis encloses the lateral walls of the Foramina of the carotid channel and of Foramen jugulare-Fossa jugularis. Fossa jugularis is bulging into Pyramis enclosing Bulbus superior venae jugularis. Processus styloideus is interposed between Fossa jugularis and Processus vaginalis. Fissura sphenopetrosa is a wide and deep rim, which is the bed of Tuba.

Course of Canalis caroticus (Figs. 42 and 43)

It starts at the Apertura int., which encloses Apex pyramidis and Apertura externa, close to Foramen lacerum. The channel makes an oblique bending into a laterodorsal direction. This segment is located close to the anterior segments of Labyrinth and Meatus acusticus int. and ext. The anterior course of the carotid channel is running almost horizontally, excentric dorsolateral located in the middle and apical segment of Pyramis. This horizontal segment is enclosed by a spongious bony bloc of Pyramis, which doesn't contain other essential structures.

Dorsal and basal axis of Pyramis
(Figs. 44, 45, 47 and 48)

The dorsal axis of pyramis is the most important for surgery. But axis and superior margin of Pyramis are incongruent. This is irrelevant for surgery if other landmarks are available. For transnasal endoscopy it may be useful to define a straight-lined basal extracranial axis along Facies inferior pyramidis. This is achieved by connecting Foramen stylomastoideum to the posterior edge of the base of Lamina medialis of Processus pterygoideus. This edge protrudes to the anterior margin of Foramen lacerum (Fig. 48). The angle of this axis to the biforaminal line (between both Foramina stylomastoidea) varies between 44° and 55° (Fig. 44). A possible asymmetry of the cranial base should be taken into consideration (see B in Fig. 44).

Cadaver skull dissections (Figs. 45 to 63)

Apex pyramidis, Facies posterior, anterior and inferior (Fig. 45)

The transectional plane of Fig. 45 presents the excentric dorsolateral position of the carotid channel and the spongious substance of the other segments of the pyramis.

Facies posterior pyramidis (Fig. 46)

Hamulus pyramidis and Impressio trigemini are located close to the apex. Fissura petrooccipitalis is small in contrast to this fissure at Facies inferior pyramidis.

Facies anterior pyramidis (Fig. 47)

Hamulus pyramidis and Impressio trigemini are located at the superior margin of Pyramis between Facies anterior and posterior.
Fisch's blue line defines Meatus acusticus int. for supratentorial surgical approaches.

Facies inferior pyramidis (Figs. 48 to 50)

From a basal perspective the projection of Canalis caroticus follows the basal axis of Pyramis. Other projections of Canalis caroticus are seen in Figs. 51 and 52.

Variants of Pyramis (Figs. 53 to 56)

Variants and normal findings may be mixed, dependent on different views (Fig. 53). Variable dorsolateral gaps of Canalis caroticus, often combined with a widening of Fissura sphenooccipitalis, are common findings. They are more relevant for supratentorial surgical approaches than for transnasal endoscopy. During supratentorial surgical approaches the bony wall of Canalis caroticus doesn't protect the artery from surgical manipulations as it does during medial basal approaches.
The bending of Canalis caroticus may be steep or flat (Fig. 55). The relationship of Labyrinth to Canalis caroticus depends on the degree of bending. Further variants are shown at the extracranial shape of the cranial base (Figs. 54 to 56). The variability in

distance and types of Spina angularis, A. meningea media and others may be relevant for defining of Apertura tubae, which is located in the depth, enclosed by the other structures.

Structures adjacent to Labyrinth (Figs. 60 to 63)

According to the well known anatomy of Pars tympanica and Labyrinth, the adjacent structures of Labyrinth are illustrated here.

Relationship of Canalis caroticus to the labyrinth bloc

Anterior area of the Labyrinth bloc (Fig. 60)
Fig. 60 presents a steep type of Canalis caroticus. Its lateral wall is crossed by Tuba. The carotid canal is located dorsal and parallel to Tuba.
A flat bending of the carotid canal results in a longer distance to Pars ossea tubae to the anterior segments of the Labyrinth (basal area of Cochlea) and to Meatus acusticus int. In this anatomical setting Cochlea and Meatus acusticus int. may be more relevant as neuronavigatory landmarks than the bending of the carotid canal.

**Area of Labyrinth, Apertura ext. can. carotici and of Fossa jugularis
(Figs. 60 to 63)**
This area is located posterior to Apertura externa of the carotid canal. Labyrinth, the base of Meatus acusticus ext. and int., and the Fallopian channel are located close to each other (Fig. 60). This short distance is further illustrated in transectional planes. Fig. 62 shows the short distance between Canalis caroticus and Cochlea. Note the short distance between Bulbus sup. v. jugularis (Fossa jugularis) and Cavum tympani at B in Fig. 61, and the discrete bulging of the wall of Cavum tympani by Processus styloideus –15- of Fig. 63.

PYRAMIS (PETROUS BONE) (Figs. 41 to 63)

Fig. 41

Overview

FIG. 41

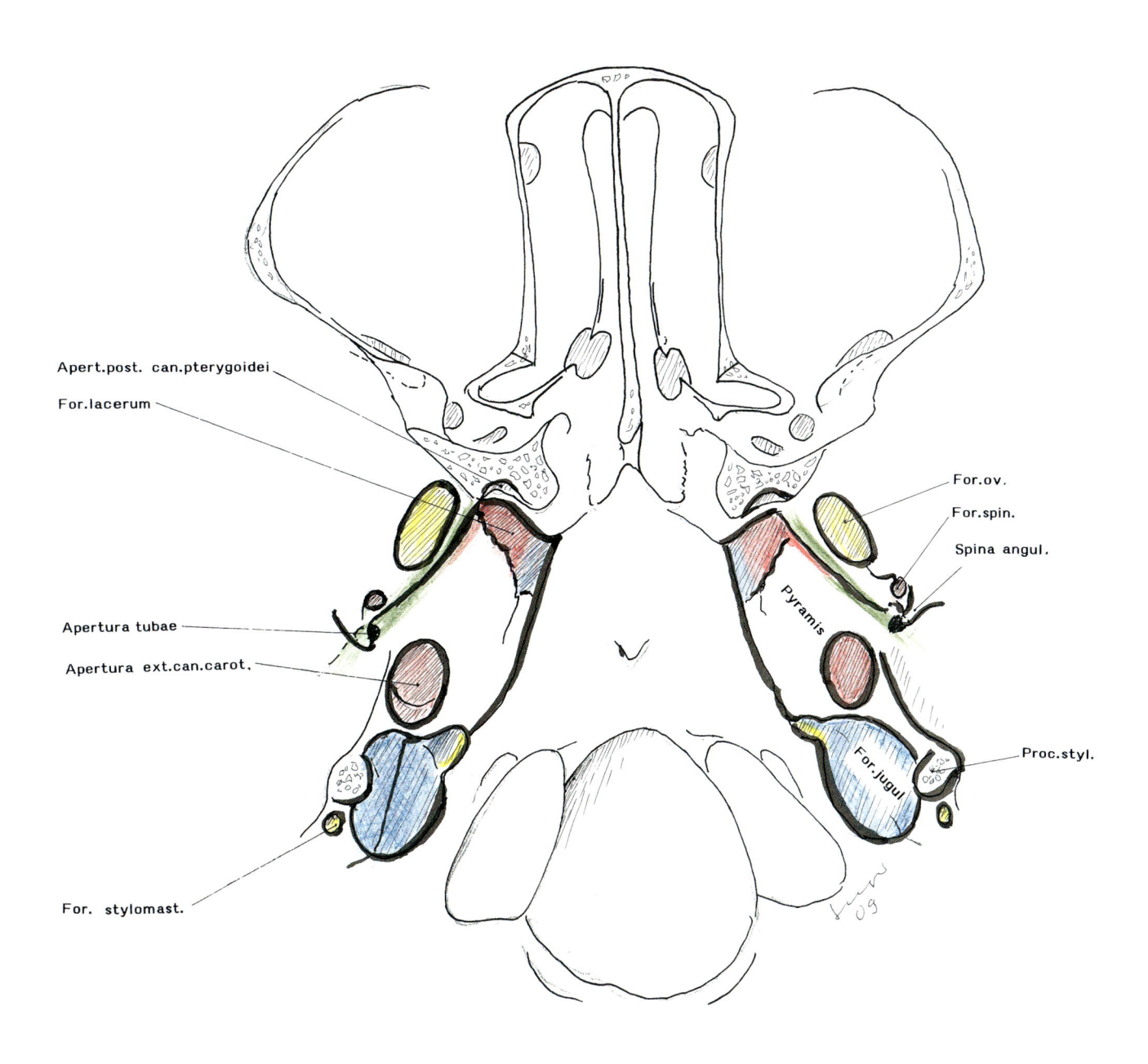

Apert.post. can.pterygoidei

For.lacerum

Apertura tubae

Apertura ext.can.carot,

For. stylomast.

For.ov.

For.spin.

Spina angul.

Pyramis

Proc.styl.

For.jugul

SIMPLIFIED PRESENTATION OF PARS PETROSA (PYRAMIS-SEGMENT)
(Figs. 42 and 43)

Fig. 42

A   Intracranial posterior shape of Pyramis. Projections of other shapes and some essential structures added

Abbreviations
1   Apex pyramidis
2a   Impressio trigemini
2b   course of Sulcus (Sinus) petrosus sup.
3   Porus acusticus int.
4   Apertura ext. ductus endolymphatici at Fossa sacculi
5   Sulcus (Sinus) sigmoideus
6   Foramen jugulare, posterior segment enclosing Bulbus sup. of V. jugularis
7   Processus intrajugularis
8   Foramen juglare, anterior segment enclosing Nn. IX to XI
9   Processus styloideus
10   Processus vaginalis of 9
11   Apertura ext. of Canalis caroticus
(11)   as 11, projection
(12)   oblique vertical segment of Canalis caroticus, projection
(13)   Curvatura of Canalis caroticus, projection
(14)   oblique horizontal segment of Canalis caroticus, projection
(15)   Area of Apertura int. of Canalis caroticus at Foramen lacerum, projection
16   Area of Sulcus (Sinus) petrosus inferior

Facies posterior pyramidis

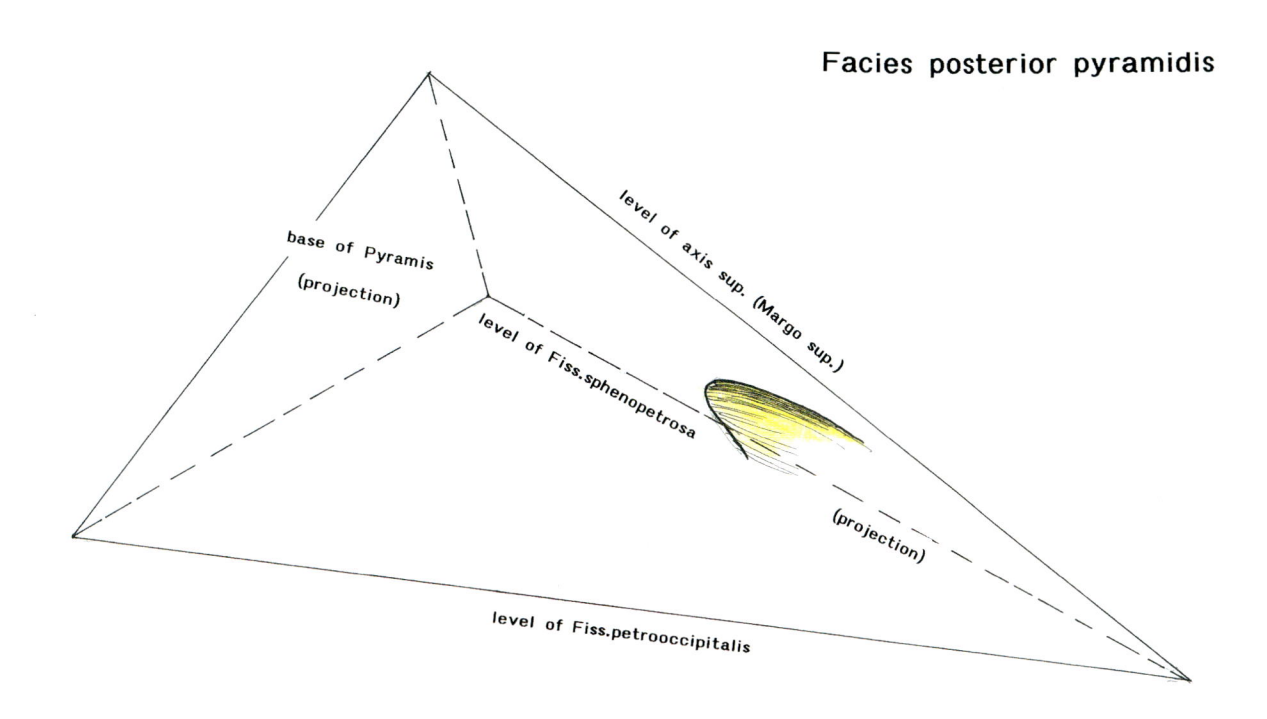

base of Pyramis
(projection)

level of axis sup. (Margo sup.)

level of Fiss.sphenopetrosa

(projection)

level of Fiss.petrooccipitalis

A

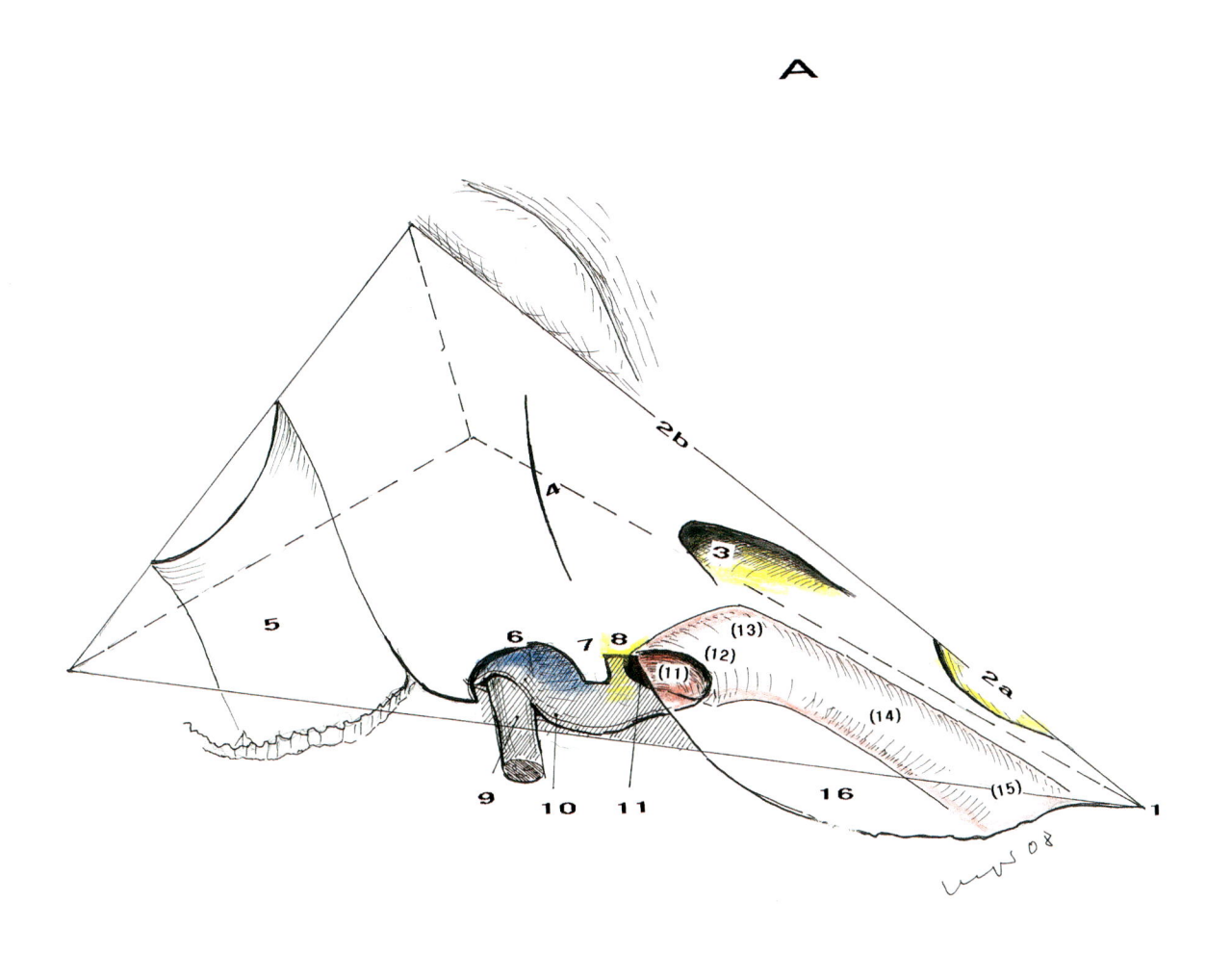

Fig. 43

Continuation of Fig. 42

B   Intracranial anterior shape of Pyramis
C   Extracranial inferior shape of Pyramis

Abbreviations
15   Area of Apertura interna canalis carotici
16   see Fig. 33
17   Foramen lacerum
18   Foramen ovale of Os sphenoidale
19   Foramen spinosum of Os sphenoidale
(20)   Apertura tubae (projection)
21   Spina angularis of Os sphenoidale
22   Eminentia arcuata (enclosing Ductus semicircularis sup., dotted)
23   Foramen stylomastoideum (enclosing N. facialis)

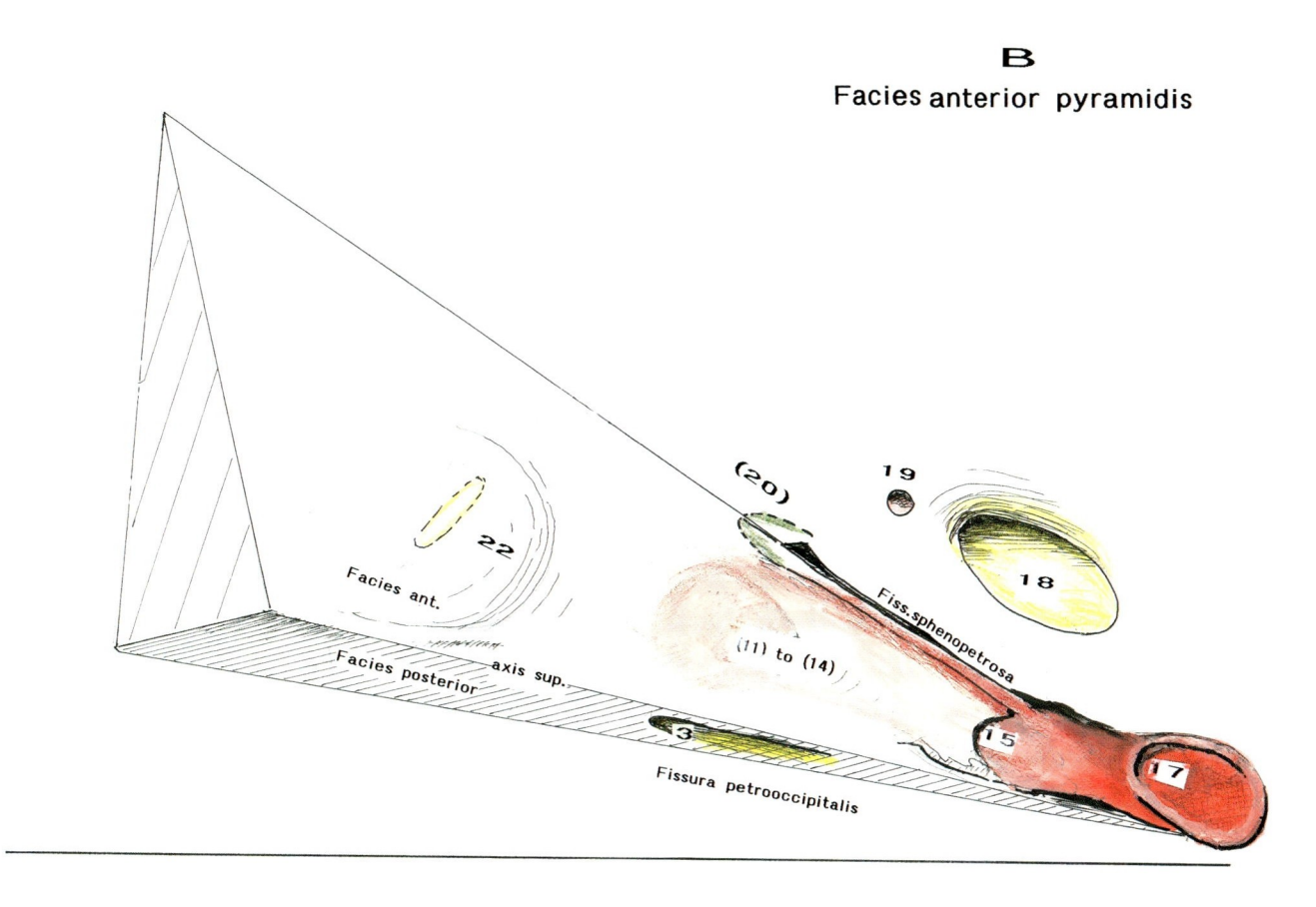

B
Facies anterior pyramidis

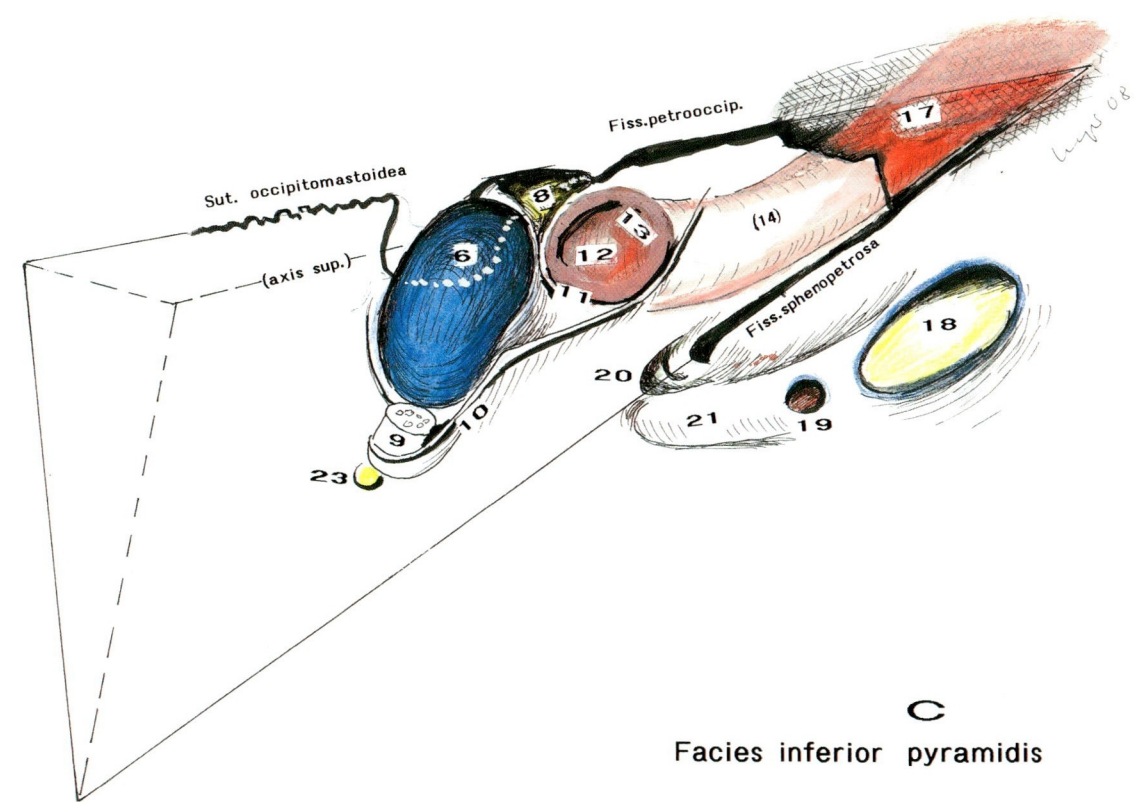

C
Facies inferior pyramidis

Fig. 44

Inferior axis of Pyramis

This axis is different from the conventional axis along the superior margin between Facies anterior and posterior pyramidis.
This inferior axis of pyramis connects Foramen stylomastoideum to the base of Lamina medialis of Processus pterygoideus. Its variations are minimal, if the connection of both Foramina stylomastoidea is positioned exactly vertical to the midline. If the cranial base is asymmetric, this should be taken into consideration (see example D)

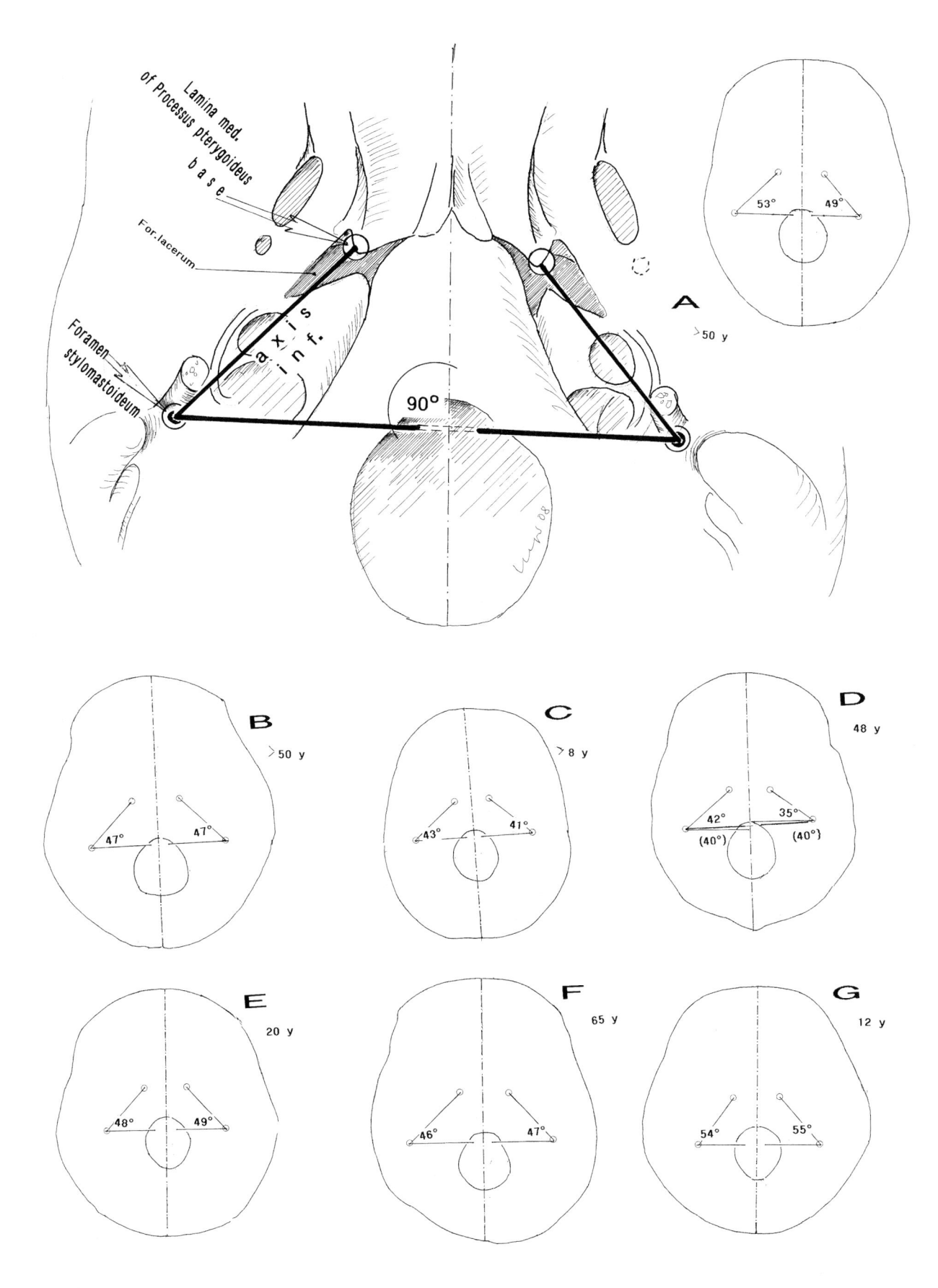

Lamina med. of Processus pterygoideus — b a s e

For.lacerum

Foramen stylomastoideum

axis inf.

90°

A
>50 y

53° 49°

B
>50 y
47° 47°

C
>8 y
43° 41°

D
48 y
42° 35°
(40°) (40°)

E
20 y
48° 49°

F
65 y
46° 47°

G
12 y
54° 55°

Fig. 45

Apex of the triangular Pyramis and Canalis caroticus

A   Topogram
B   Pyramis and adjacent structures

Abbreviations
1   Apex pyramidis
2   transectional plane
3   A. carotis int.
4   Tuba auditiva (Eustachii)
5   A. meningea media
6   Foramen ovale
7   defect and thin wall of Canalis caroticus
8   defect and thin wall of the ground of Fissura sphenopetrosa

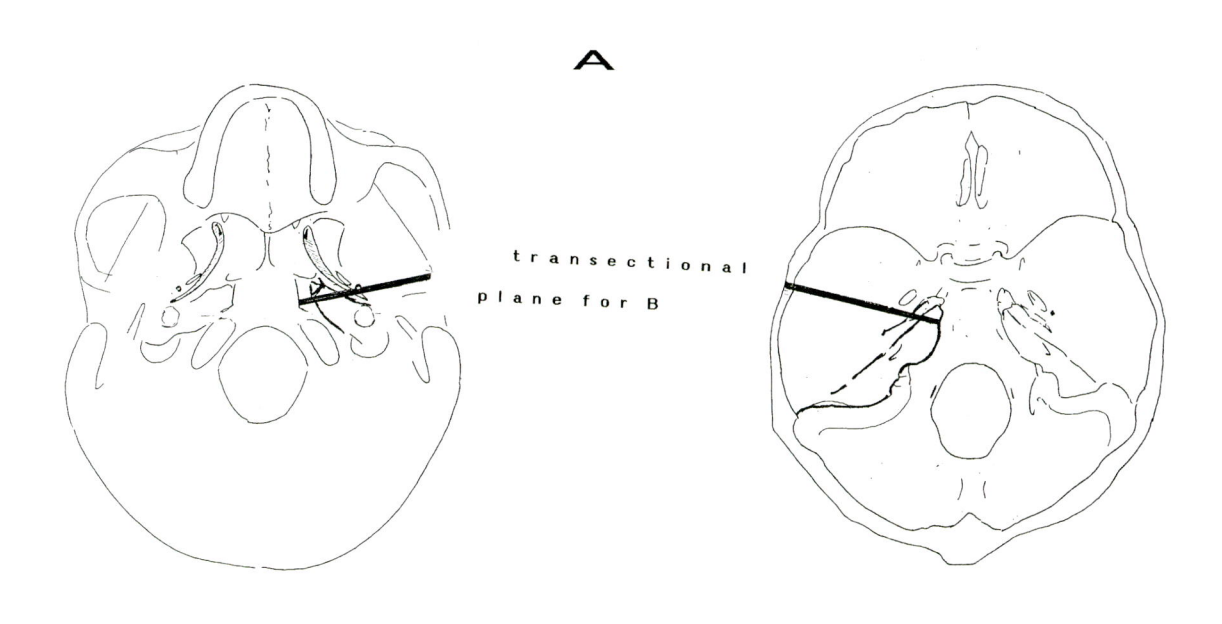

A

transectional
plane for B

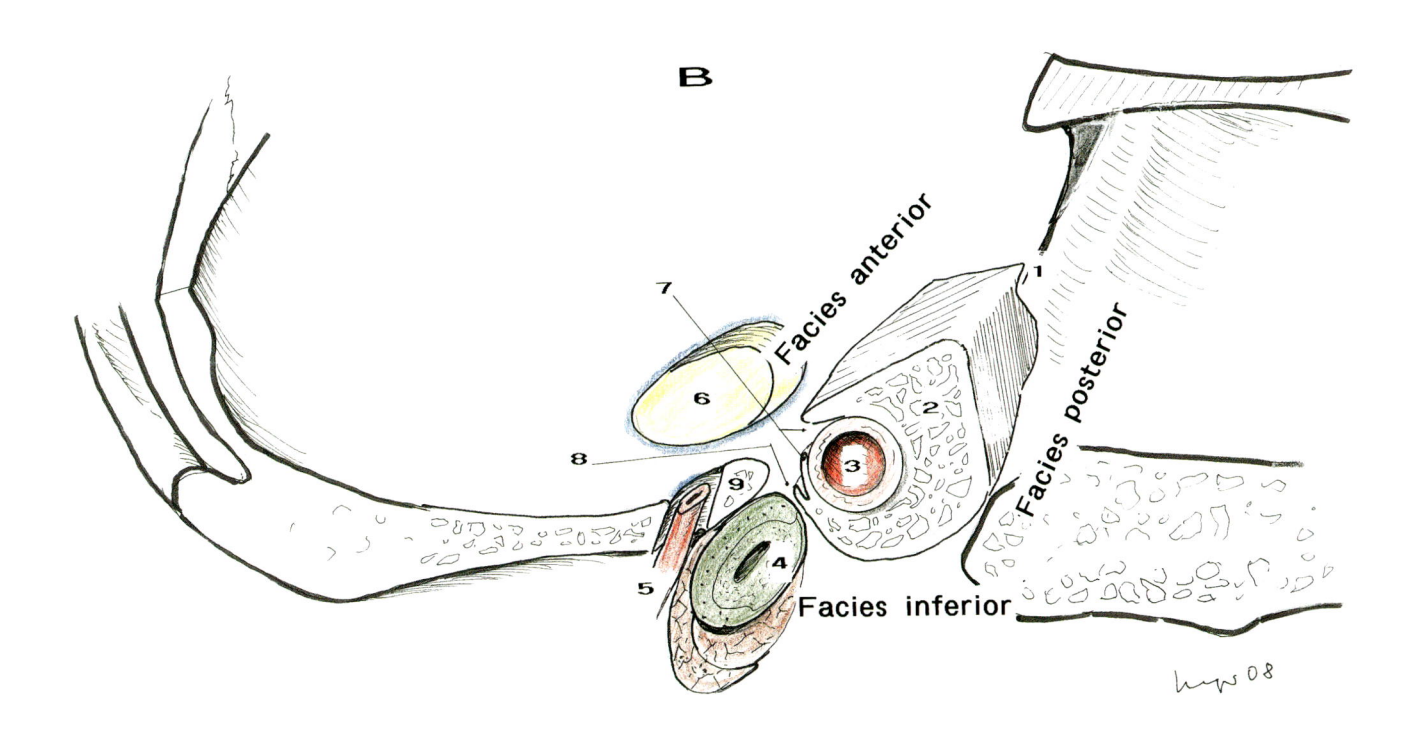

B

Facies anterior

Facies posterior

Facies inferior

Fig. 46

Facies posterior pyramidis and surrounding structures.
Labyrinth, Meatus acusticus int. and Canalis caroticus transparent or dotted.

Abbreviations
1   Impressio trigemini
2   Sulcus petrosus sup.
3   Eminentia arcuata
4   Porus acusticus int.
(5)   Fundus of Meatus acusticus int.
(6)   Cochlea, inferior margin

FIG. 46

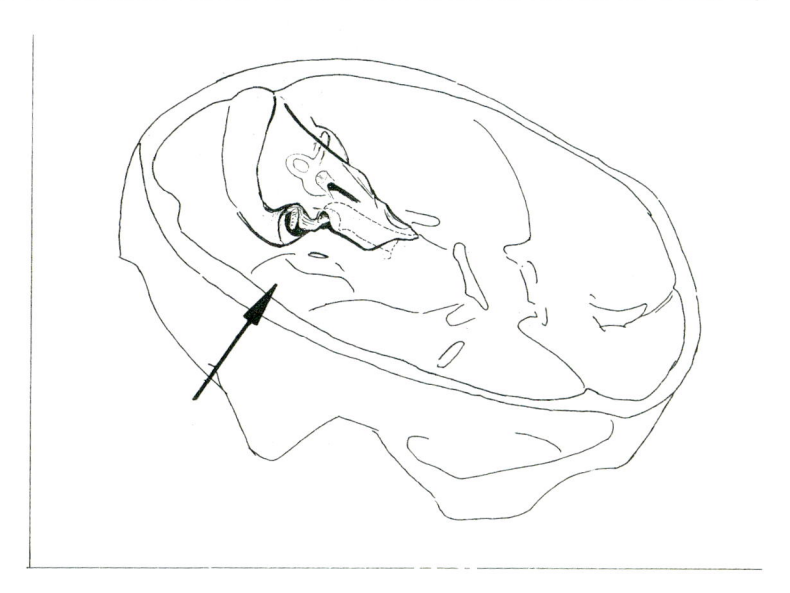

Facies posterior pyramidis

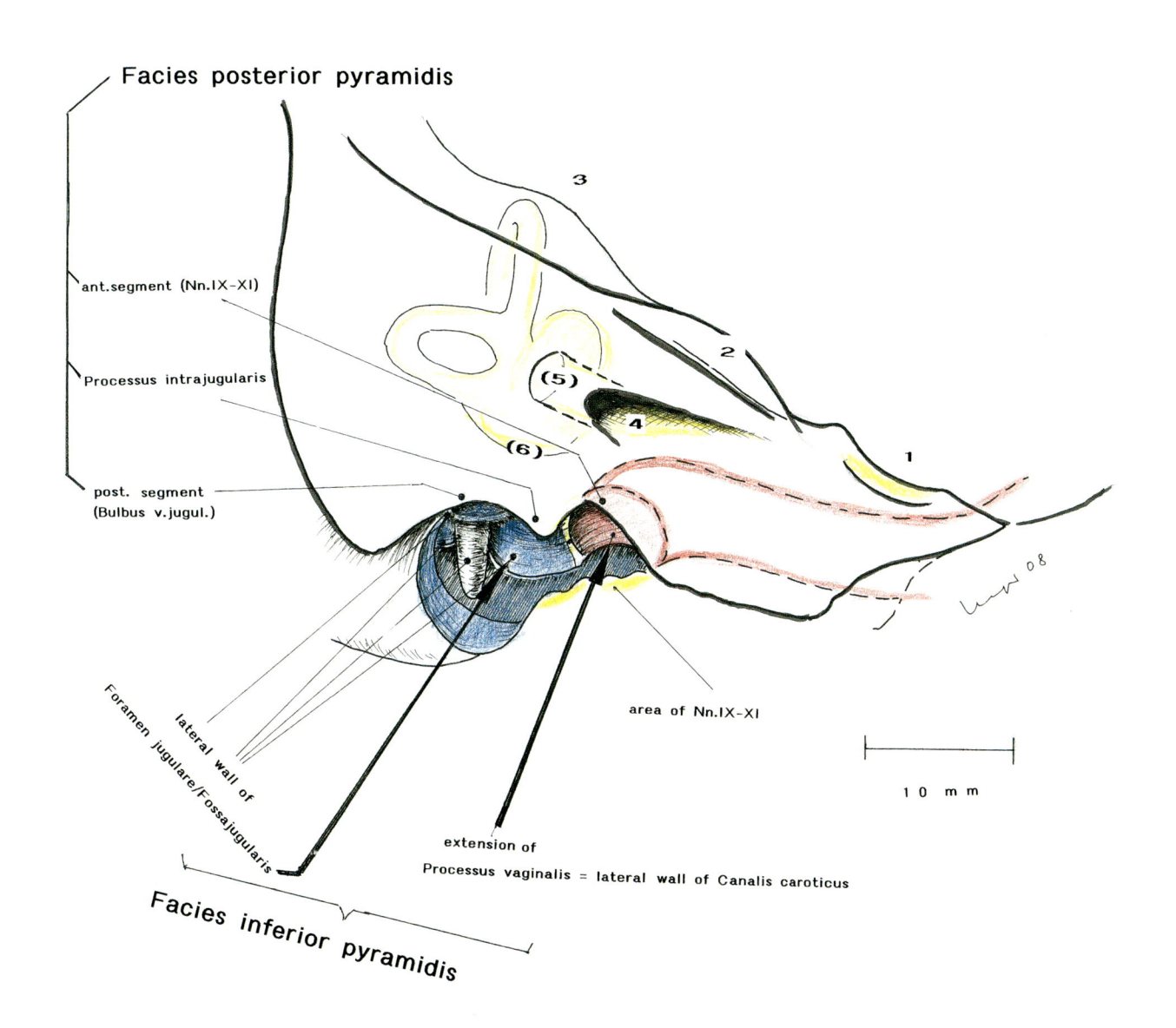

3

ant.segment (Nn.IX–XI)

Processus intrajugularis

(5)

(6)

2

4

1

post. segment
(Bulbus v.jugul.)

area of Nn.IX–XI

Foramen jugulare/Fossa jugularis

lateral wall of

extension of
Processus vaginalis = lateral wall of Canalis caroticus

10 m m

Facies inferior pyramidis

Fig. 47

Facies anterior and posterior pyramidis and surrounding structures

Contents and dorsal axis added.
The axis is different from the superior margin of Pyramis
Oblique projection of Facies posterior pyramidis
For blue line see text.

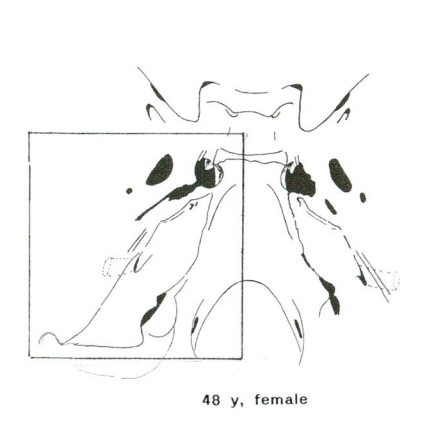

48 y, female

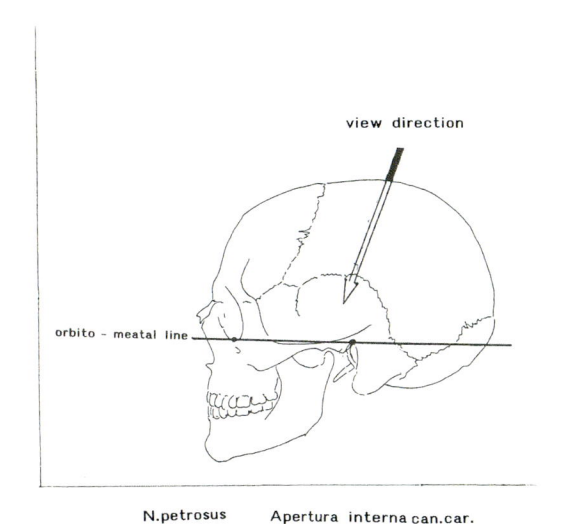

view direction

orbito – meatal line

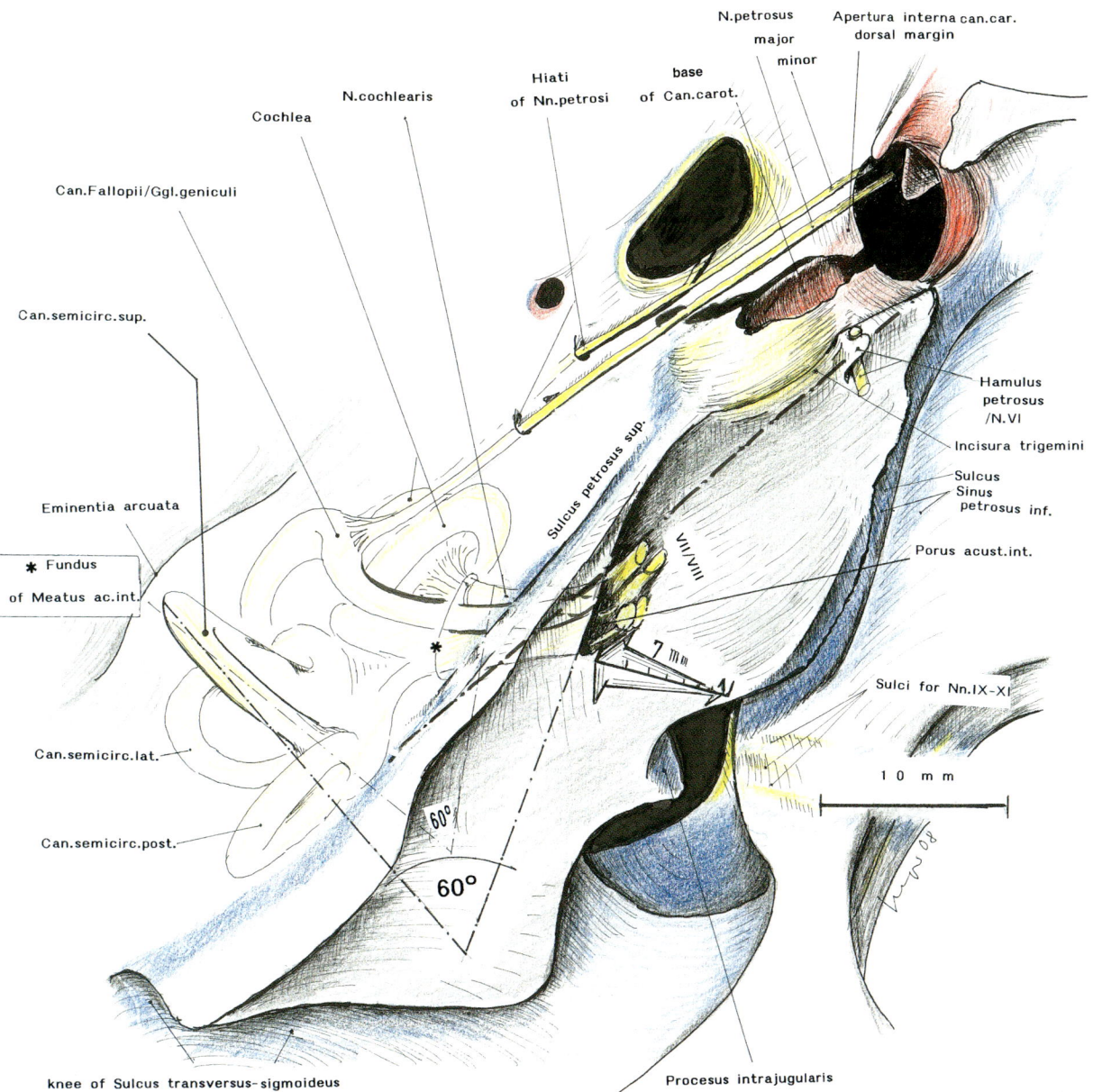

Can.Fallopii/Ggl.geniculi

Cochlea

N.cochlearis

Hiati
of Nn.petrosi

N.petrosus
major
minor

base
of Can.carot.

Apertura interna can.car.
dorsal margin

Can.semicirc.sup.

Hamulus
petrosus
/N.VI

Incisura trigemini

Sulcus
Sinus
petrosus inf.

Eminentia arcuata

Sulcus petrosus sup.

VII/VIII

Porus acust.int.

* Fundus
of Meatus ac.int.

7 mm

Sulci for Nn.IX-XI

Can.semicirc.lat.

10 mm

Can.semicirc.post.

60°

60°

knee of Sulcus transversus-sigmoideus

Procesus intrajugularis

Fig. 48

Facies inferior pyramidis and surrounding structures.
Inferior axis added

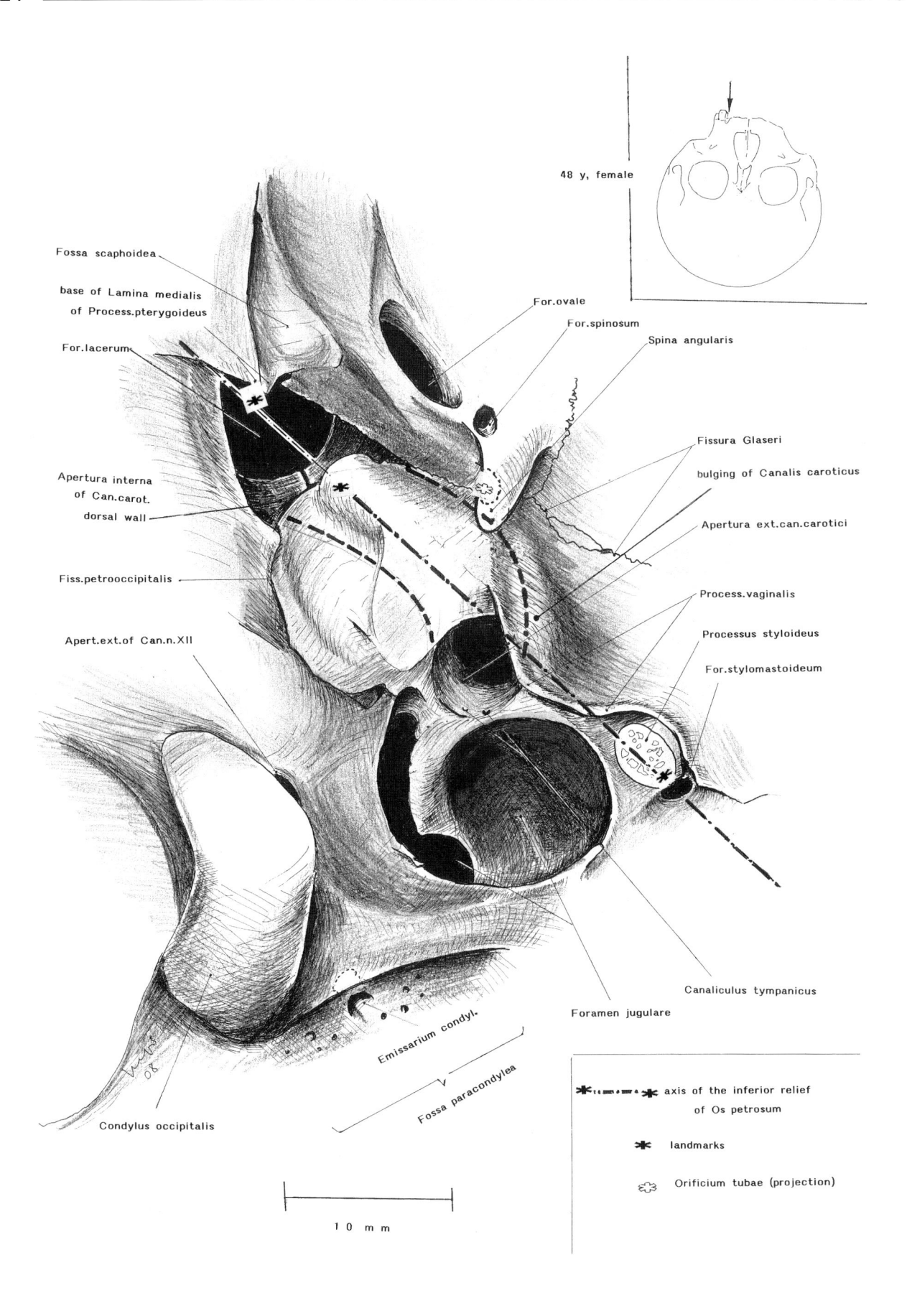

48 y, female

Fossa scaphoidea

base of Lamina medialis
of Process.pterygoideus

For.lacerum

Apertura interna
of Can.carot.
dorsal wall

Fiss.petrooccipitalis

Apert.ext.of Can.n.XII

Condylus occipitalis

Emissarium condyl.

Fossa paracondylea

For.ovale

For.spinosum

Spina angularis

Fissura Glaseri

bulging of Canalis caroticus

Apertura ext.can.carotici

Process.vaginalis

Processus styloideus

For.stylomastoideum

Canaliculus tympanicus

Foramen jugulare

10 mm

✳ ··─··─·· ✳ axis of the inferior relief
of Os petrosum

✳ landmarks

Orificium tubae (projection)

Fig. 49

Addendum to Fig. 48
Nerves and blood vessels added

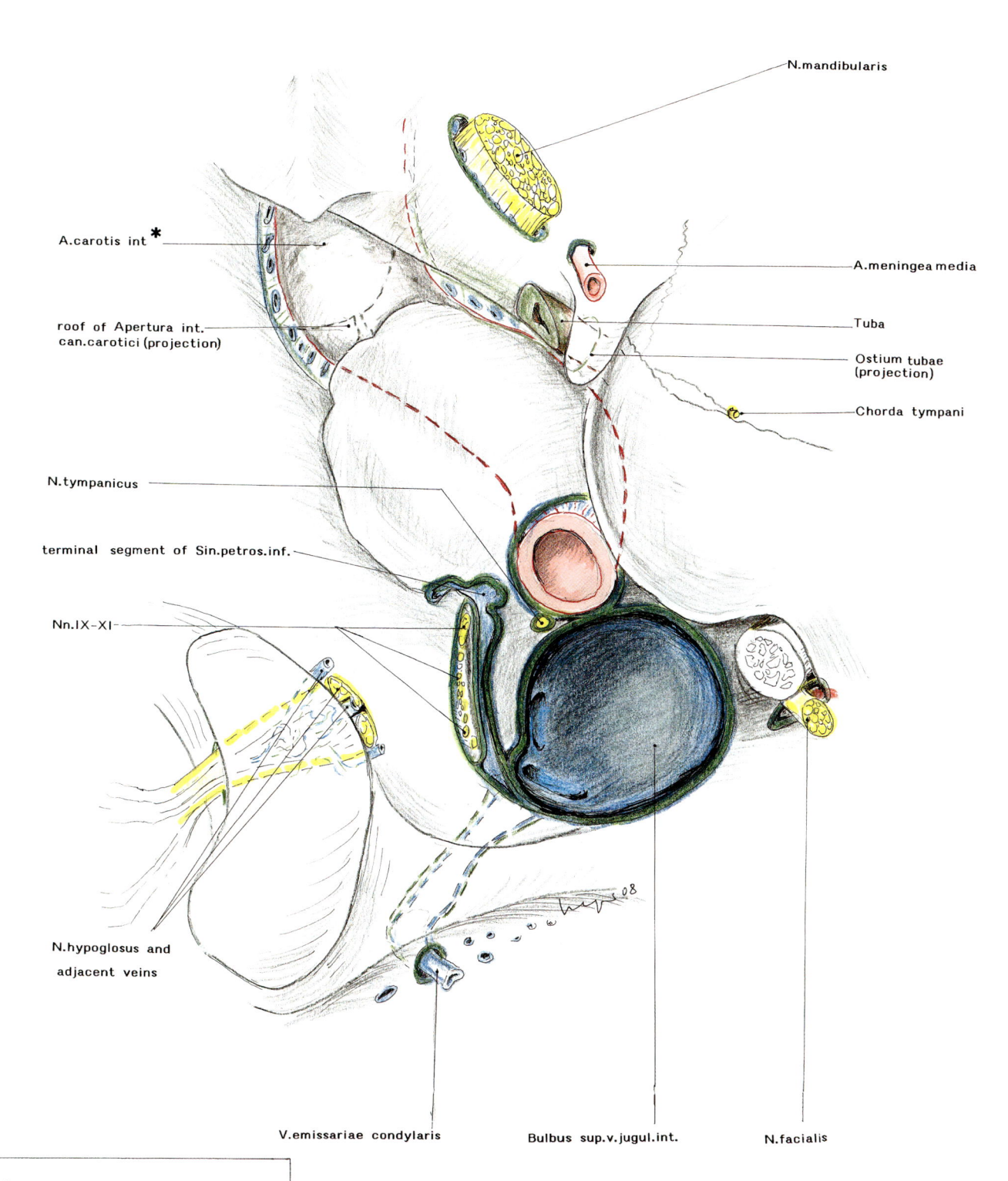

N.mandibularis

A.carotis int *

A.meningea media

roof of Apertura int.
can.carotici (projection)

Tuba

Ostium tubae
(projection)

Chorda tympani

N.tympanicus

terminal segment of Sin.petros.inf.

Nn.IX–XI

N.hypoglosus and
adjacent veins

V.emissariae condylaris

Bulbus sup.v.jugul.int.

N.facialis

* Enchondrosis sphenopetrosa omitted

Fig. 50

Facies inferior pyramidis, posterior segment

Note the short distance between all segments.

A Apertura tubae and Fissura/Sutura sphenopetrosa (bed of Tuba)
B As A, sectional enlargement
C Tuba and its adjacent muscles

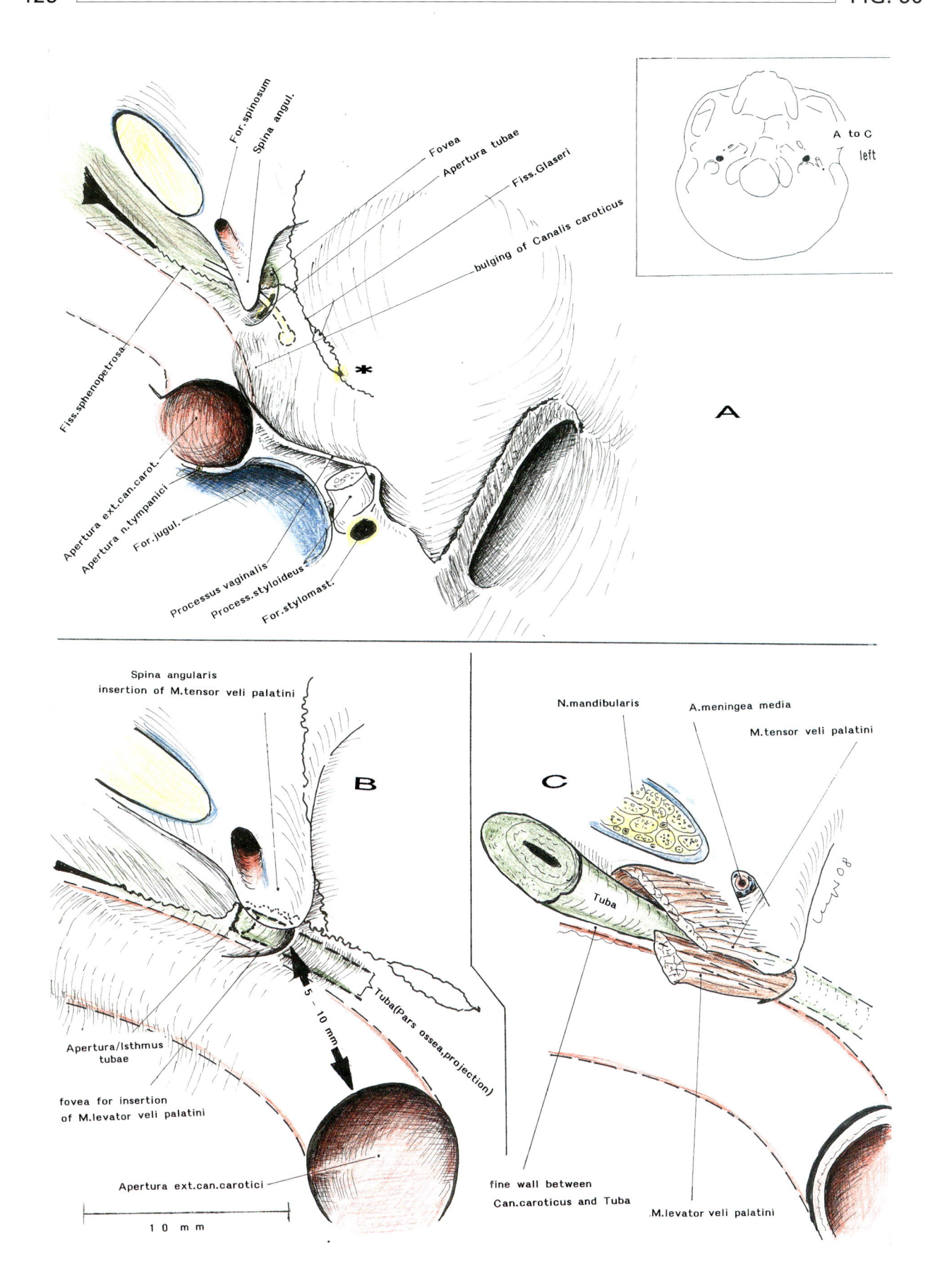

For.spinosum
Spina angul.
Fovea
Apertura tubae
Fiss.Glaseri
bulging of Canalis caroticus
Fiss.sphenopetrosa
Apertura ext.can.carot.
Apertura n.tympanici
For.jugul.
Processus vaginalis
Process.styloideus
For.stylomast.

A to C
left

A

Spina angularis
insertion of M.tensor veli palatini

B

N.mandibularis
A.meningea media
M.tensor veli palatini

C

Tuba

5 - 10 mm

Tuba(Pars ossea,projection)

Apertura/Isthmus
tubae

fovea for insertion
of M.levator veli palatini

Apertura ext.can.carotici

10 m m

fine wall between
Can.caroticus and Tuba

.M.levator veli palatini

VARIABLE VIEW. A CAROTIS INT. ADDED (PROJECTIONS)
(Figs. 51 and 52)

Fig. 51

Abbreviations
1 Fossa paracondyloidea
2 Fossa biventerica
3 margin of Foramen jugulare
4 Foramen stylomastoideum
(4) projection
5 Processus styloideus
6 Processus vaginalis
7 Apertura externa canalis carotici
8 Fossa mandibularis
9 Apertura of Tuba to Semicanalis musculotubarius
10 Spina angularis
11 Foramen spinosum
12 Tuberculum articulare
13 Sulcus tubarius at Fissura sphenopetrosa
14 Foramen lacerum
15 outer surface of the wall of Canalis caroticus
16 Os basilare, medial margin of Foramen lacerum
17 base of Lamina medialis processi pterygoidei
18 Alae vomeris
19 Apertura externa canalis nervi hypoglossi
20 Apertura interna canalis nervi hypoglossi
21 Foramen ovale
22 Lamina lateralis processi pterygoidei

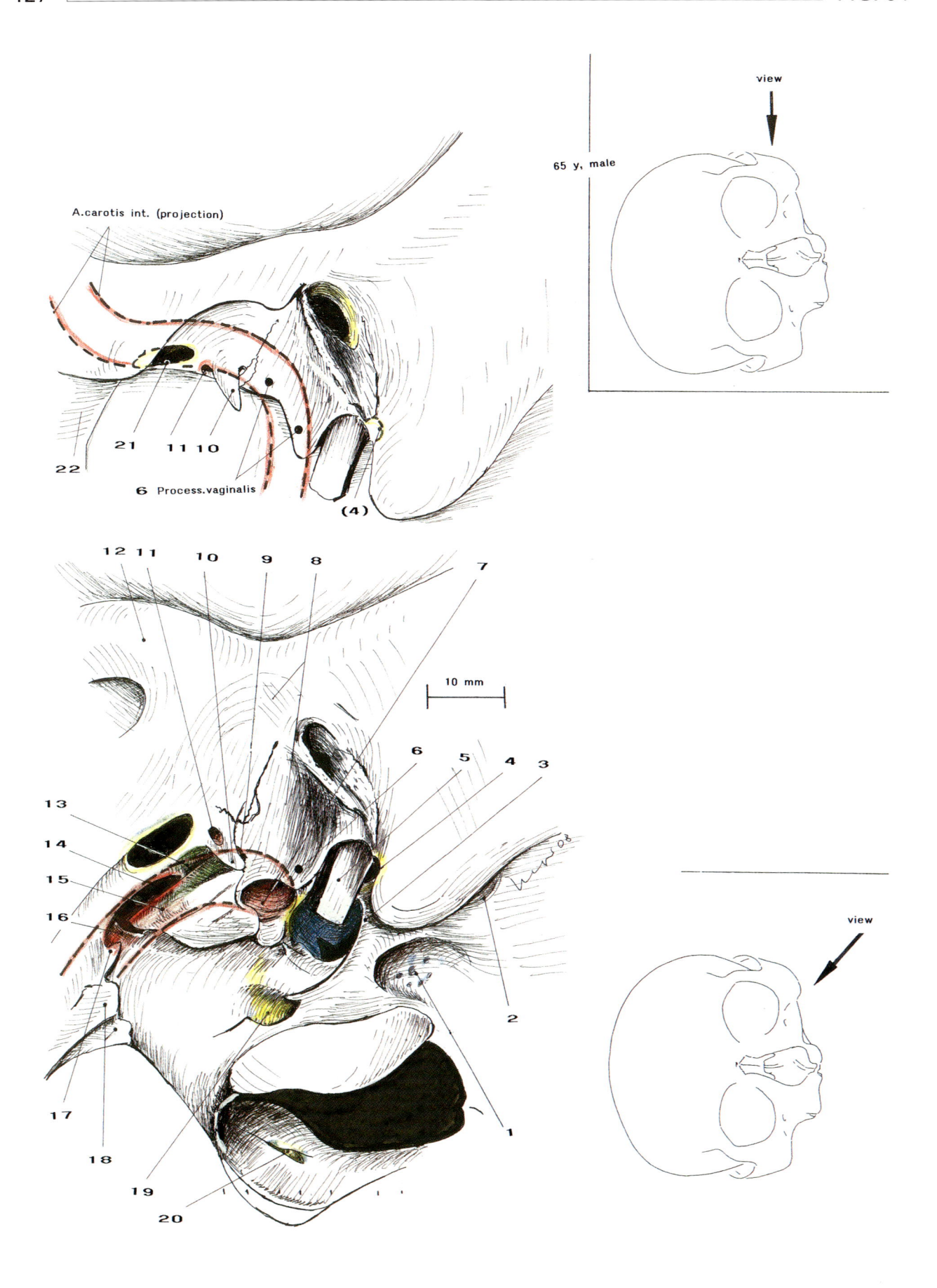

A.carotis int. (projection)

21 11 10

22

6 Process.vaginalis

(4)

65 y, male

view

12 11   10   9   8

7

10 mm

13

14

15

16

6   5   4   3

17

18

19

20

2

1

view

Fig. 52

Oblique view from the contralateral side
Condylus occipitalis overlaps Apertura ext. of Canalis n. XII

Abbreviations
1 Apertura interna nervi hypoglossi
2 Condylus occipitalis
3 Emissarium condylare
4 Foramen jugulare
5 Processus styloideus
6 Foramen stylomastoideum
7 Processus vaginalis
8 Processus mastoideus
9 Porus acusticus externus
10 Fissura petrotympanica Glaseri (contains Chorda tympani)
11 Sutura sphenosquamosa
12 Spina angularis
13 Foramen spinosum
14 Foramen ovale
15 Foramen lacerum
16 gap for Synchondrosis sphenooccipitalis at infancy
17 Corpus sphenoidale
18 Pars basilaris ossis occipitalis

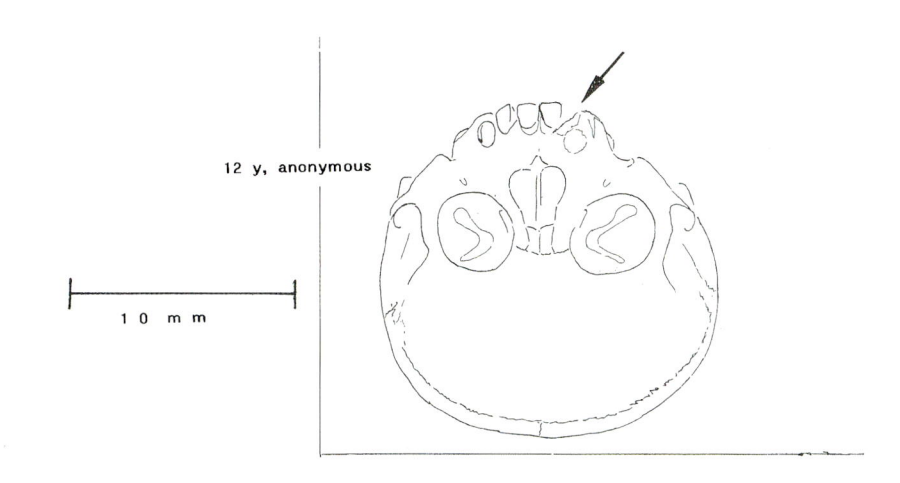

12 y, anonymous

10 mm

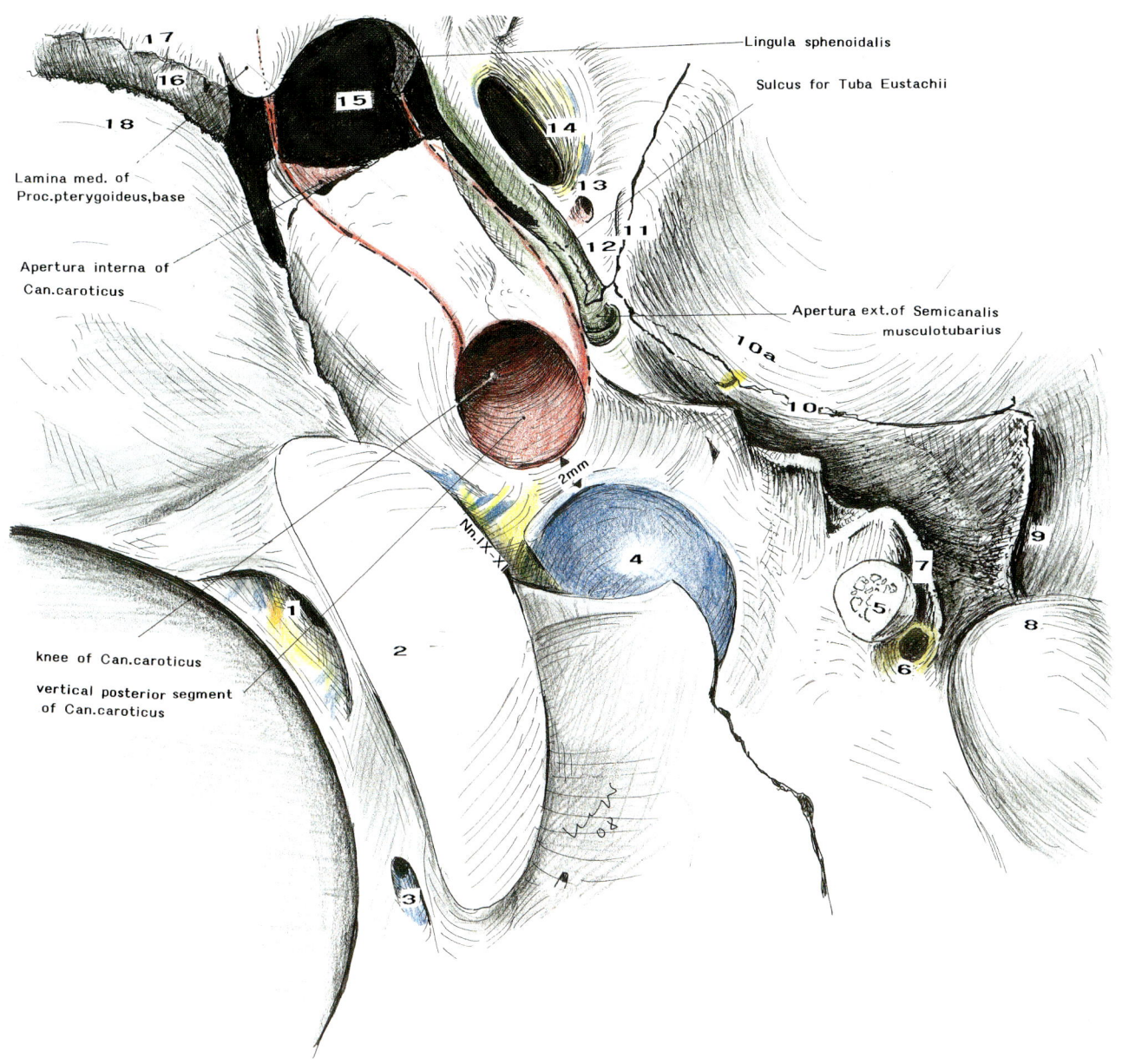

Lingula sphenoidalis

Sulcus for Tuba Eustachii

15

14

13

11

12

Lamina med. of
Proc.pterygoideus,base

Apertura interna of
Can.caroticus

Apertura ext.of Semicanalis
musculotubarius

10a

10c

2mm

Nn.IX-XI

4

9

7

5

knee of Can.caroticus

vertical posterior segment
of Can.caroticus

1

6

8

2

3

17

16

18

Fig. 53

Variant and possible errors depending on the perspective
Illustrated by cadaver skull dissections

A Normal finding
A' As A, oblique view
B Variant. View as A

Abbreviations
1 Processus mastoideus
2 Foramen stylomastoideum
3 Processus styloideus
4 Processus vaginalis
5 Apertura ext. canalis carotici
6 Foramen jugulare
7 Apertura of Pars ossea tubae (schematic)
8 Fissura (Synchondrosis) sphenopetrosa
9 Spina angularis
10 Foramen spinosum
11 Foramen ovale
12 Fissura (Synchondrosis) petrooccipitalis (schematic)
13 Canaliculus tympanicus
14 Processus intrajugularis

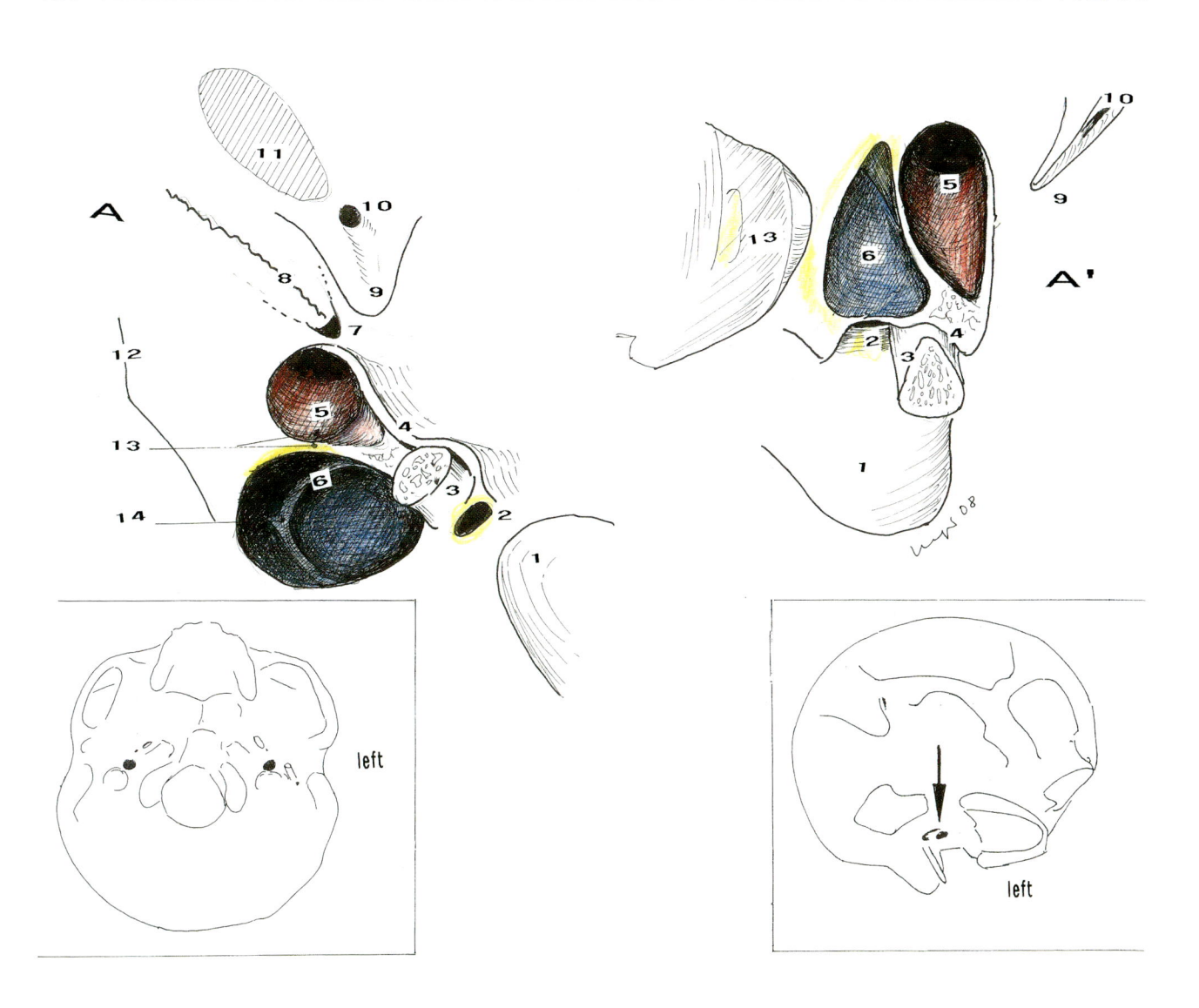

A

A'

left

left

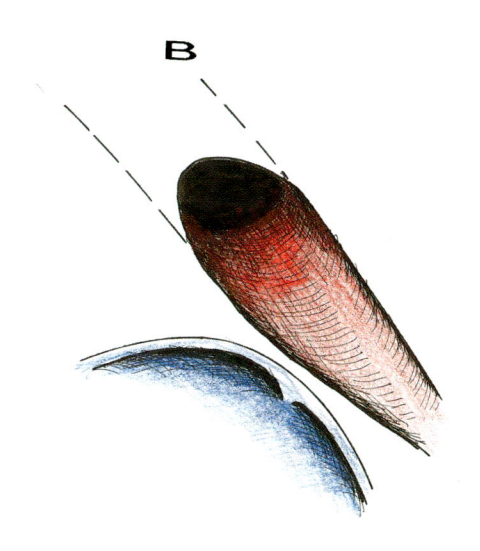

B

Fig. 54 and Fig. 55

Variants between Facies inferior pyramidis and Ala major

Foramen ovale
    exostoses (see 3 in A and B)
Foramen spinosum
    communicating to Fissura sphenopetrosa (4 at B)
Spina angularis
    Hypoplasia (B, D, and E)
Apertura tubae
    overlapped by Spina angularis: arrow in C and (14) in E
    overlapped by an exostosis of Pyramis (B)
    no overlap (arrow in A, 14 in D)
Apertura externa canalis carotici
    long distance to Apertura tubae (light arrows in B)
    short distance to Apertura tubae in C and E
Processus vaginalis
    lateral to Processus styloideus, common finding (13 in D)
    medially to Processus styloideus (rare variant, 13 in E)
    small overlap of Apertura canalis carotici (13 in E)
    wide overlap of Apertura externa canalis carotici (13 in E')
Bending of Canalis caroticus
    steep course, common finding (D')
    flat, rare variant (E and E')

Abbreviations
1   Glaser's fissure
2   Spina angularis
3   Exostosis
4   Sutura/Fissura sphenopetrosa
5   Foramen spinosum (5) absent
6   Foramen ovale
7   Emissaria
8   Apertura externa canalis carotici
9   Processus styloideus
10   Foramen stylomastoideum
11   Foramen jugulare
12   bony fovea
13   Processus vaginalis
14   Apertura tubae
(14) overlap

FIG. 54

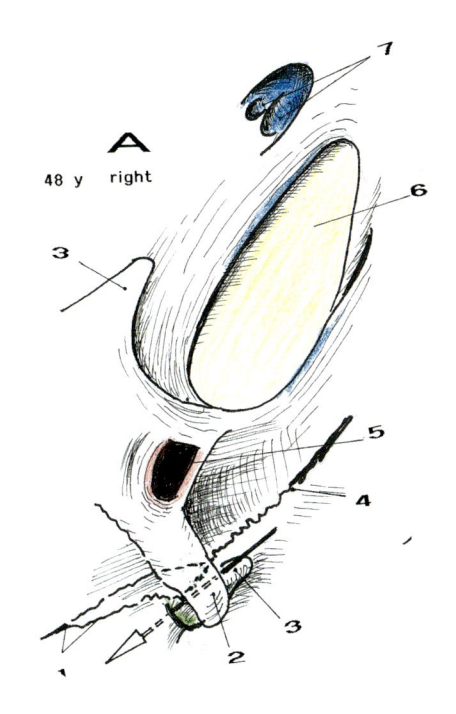

A

48 y   right

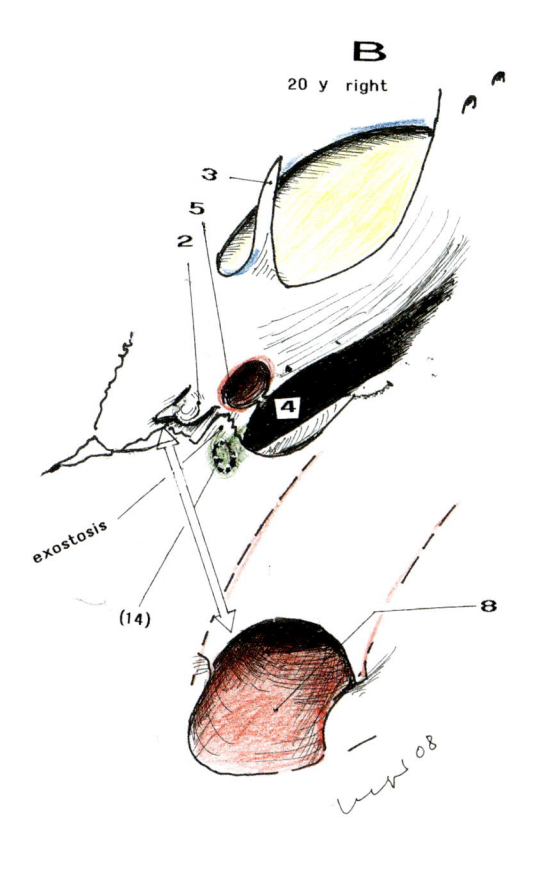

B

20 y   right

exostosis

(14)

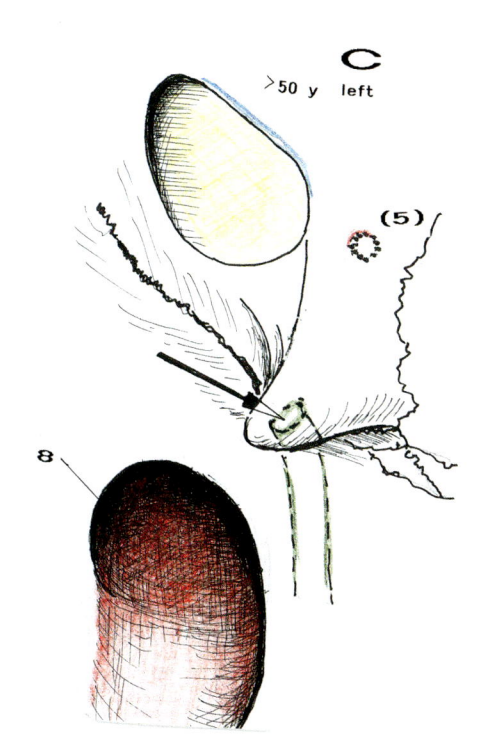

C

>50 y   left

(5)

Fig. 54 and Fig. 55

Variants between Facies inferior pyramidis and Ala major

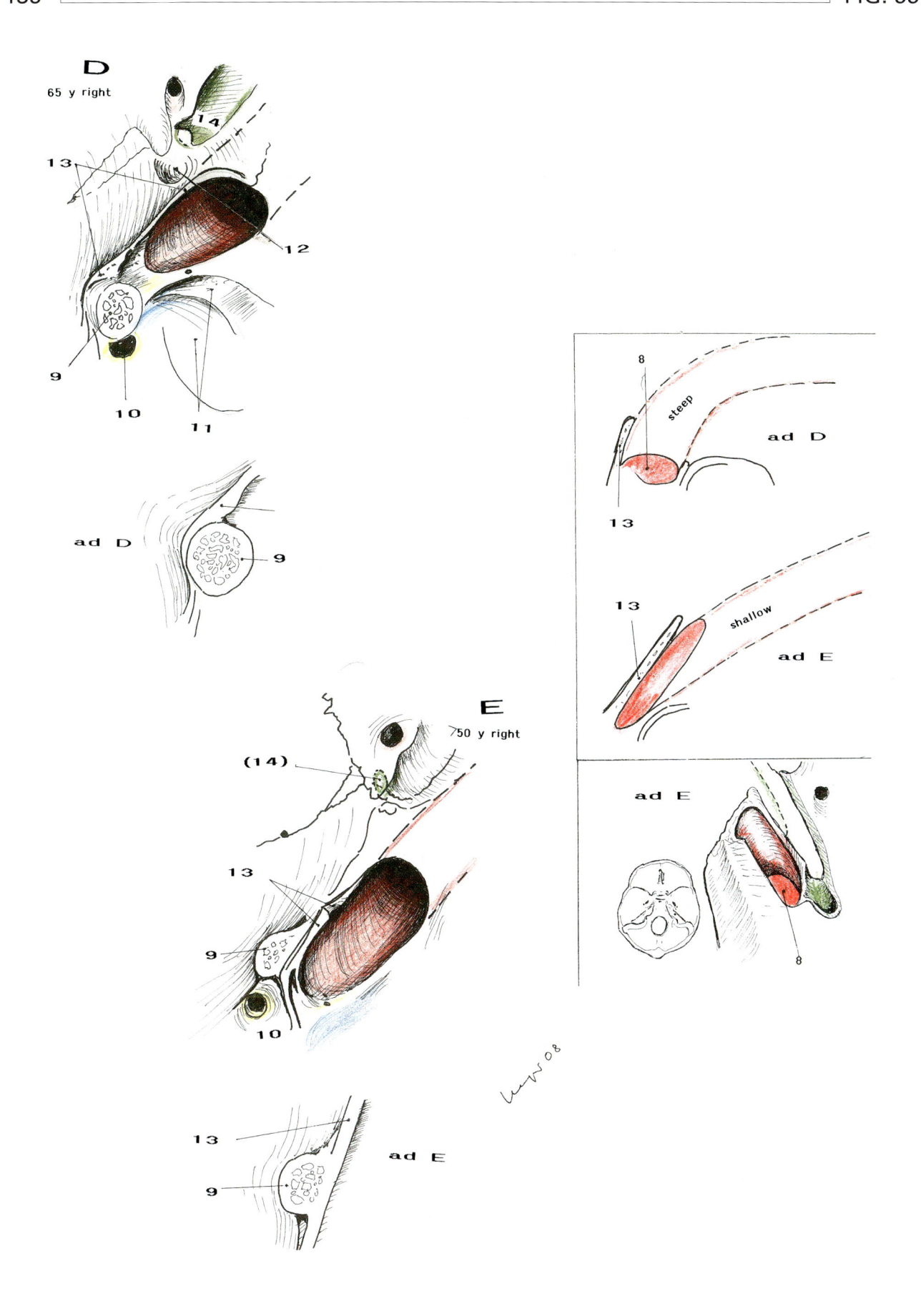

Fig. 56

Addendum to Figs. 54 and 55
Area of Spina angularis and Processus vaginalis

A   Spina angularis singular, merged with an exostosis of Pyramis -5a-,
Both overlap Apertura tubae. Arrow: Course of Tuba.
Processus vaginalis overlaps Apertura ext. of Can. caroticus
B   Doubled Spina angularis. Both components overlap Apertura tubae
C   Elongated Spina angularis. Partial overlap of Apertura tubae by an elongated
Processus vaginalis
D   Spina angularis similar to C. Overlapping Foramen spinosum
Note the difference of Processus vaginalis in C and D

Processus vaginalis was called "Vagina of Processus styloideus" in the past. But it encloses not only the lateral circumference of Apertura int. of Apertura ext. of the carotid channel. Today it is called Processus vaginalis

Abbreviations
1   Foramen ovale
2   Foramen spinosum
3   Sutura (Fissura) sphenopetrosa
4   bony Fovea
5a) Exostosis of Pyramis
5b) Exostosis of Ala major
6   Apertura ext. canalis carotici
7   Fissura Glaseri
8   Foramen stylomastoideum
9   Processus styloideus

---

* arrow: Course of Tuba

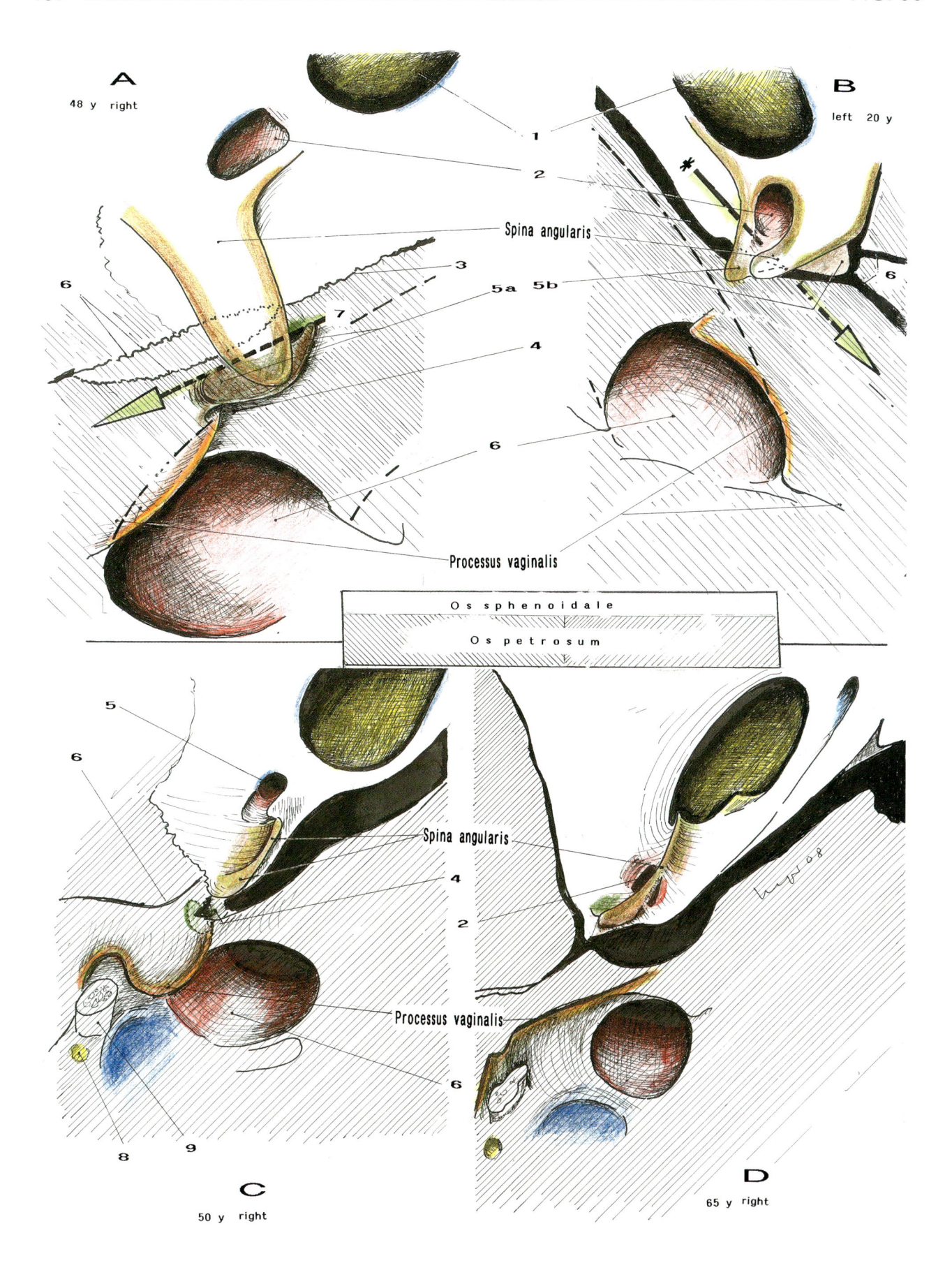

A

48 y right

Spina angularis

6

3

5a 5b

7

4

6

Processus vaginalis

B

left 20 y

*

6

6

| Os sphenoidale |
| Os petrosum |

C

5

6

Spina angularis

4

Processus vaginalis

6

8 9

50 y right

D

Spina angularis

2

4

Processus vaginalis

6

65 y right

Fig. 57

Transection of the Pyramis at the bending segment of Canalis caroticus. Its relationship to Tuba auditiva.

A　Posterior part of the dissection. The highest point of the carotid canal is not yet reached

B　Anterior part of the dissection. Relationship of Canalis caroticus and Apertura tuba marked by dots.

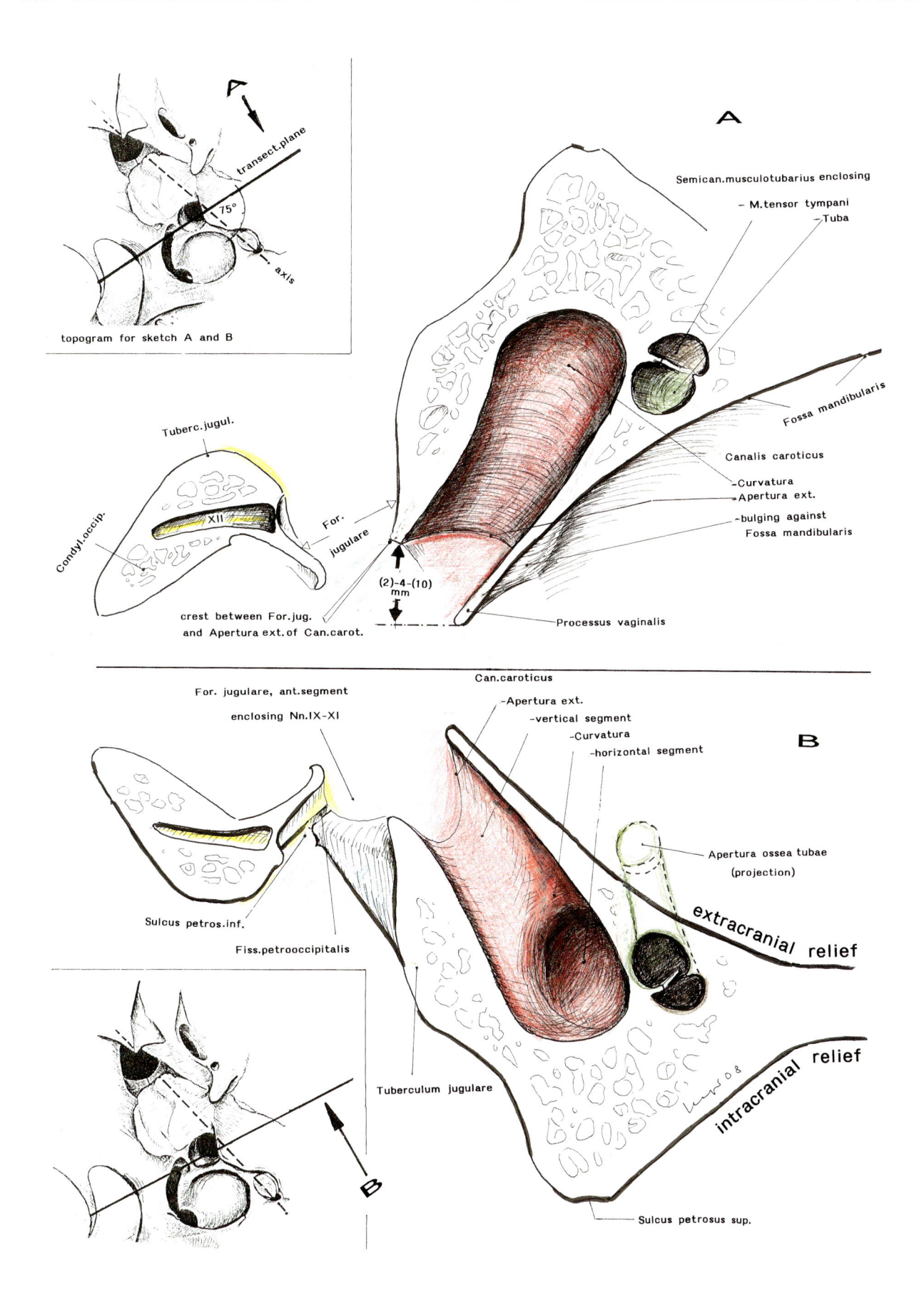

A

topogram for sketch A and B

transect.plane

75°

axis

Semican.musculotubarius enclosing
- M.tensor tympani
- Tuba

Fossa mandibularis

Canalis caroticus
- Curvatura
- Apertura ext.
- bulging against
Fossa mandibularis

Processus vaginalis

Tuberc.jugul.

Condyl.occip.

XII

For. jugulare

(2)-4-(10) mm

crest between For.jug.
and Apertura ext.of Can.carot.

Can.caroticus

For. jugulare, ant.segment
enclosing Nn.IX-XI

-Apertura ext.
-vertical segment
-Curvatura
-horizontal segment

B

Apertura ossea tubae
(projection)

extracranial relief

intracranial relief

Sulcus petros.inf.

Fiss.petrooccipitalis

Tuberculum jugulare

Sulcus petrosus sup.

Fig. 58

Horizontal segment of Canalis caroticus

A Transectional level of B (arrows). Canalis caroticus dotted.
B Transectional plane of Apex pyramidis, enlarged
Thin walled dorsolateral segment of Canalis caroticus resected

Abbreviations
1 Fossa jugularis
2 Foramen stylomastoideum
3 as 5
4 Processus styloideus, variant
5 Processus vaginalis
6 Fissura Glaseri
7a) Canalis caroticus
  b) Canalis caroticus, distal (horizontal) segment
8 Apertura tubae
9 Fissura sphenopetrosa
10 Spina angularis
11 Foramen spinosum
12 Foramen ovale
13 Canalis caroticus, close to Apertura int.
14 base of Lamina medialis processi pterygoidei
15 Foramen lacerum
16 Tuberculum pharyngeum
17 Fissura petrooccipitalis
18 Canalis nervi hypoglossi

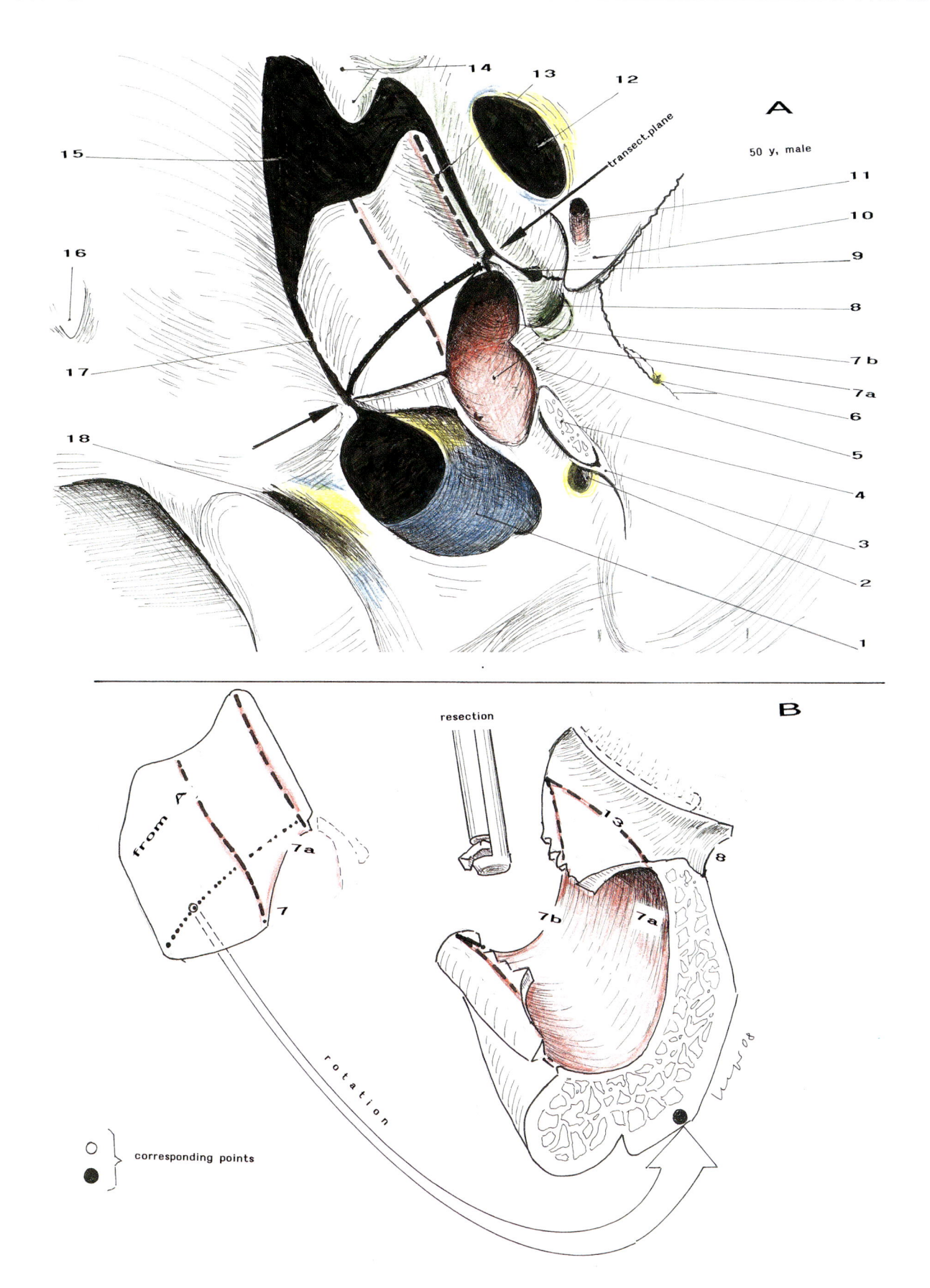

A

50 y, male

transect.plane

resection

from A

7a

7

rotation

corresponding points

B

13

8

7b

7a

Fig. 59

Horizontal segment of Canalis caroticus, variants

A and **B** Closed dorsal wall at Apertura int. canalis carotici. Common variant
C   Widening of Fissura sphenopetrosa. Foramen spinosum enclosed by the fissure
D   Usual findings at Canalis caroticus and Fissura sphenopetrosa.
Foramen spinosum not shown (Lang, 1979) For. surgical aspects see Seeger, 2003, pp 112 ff

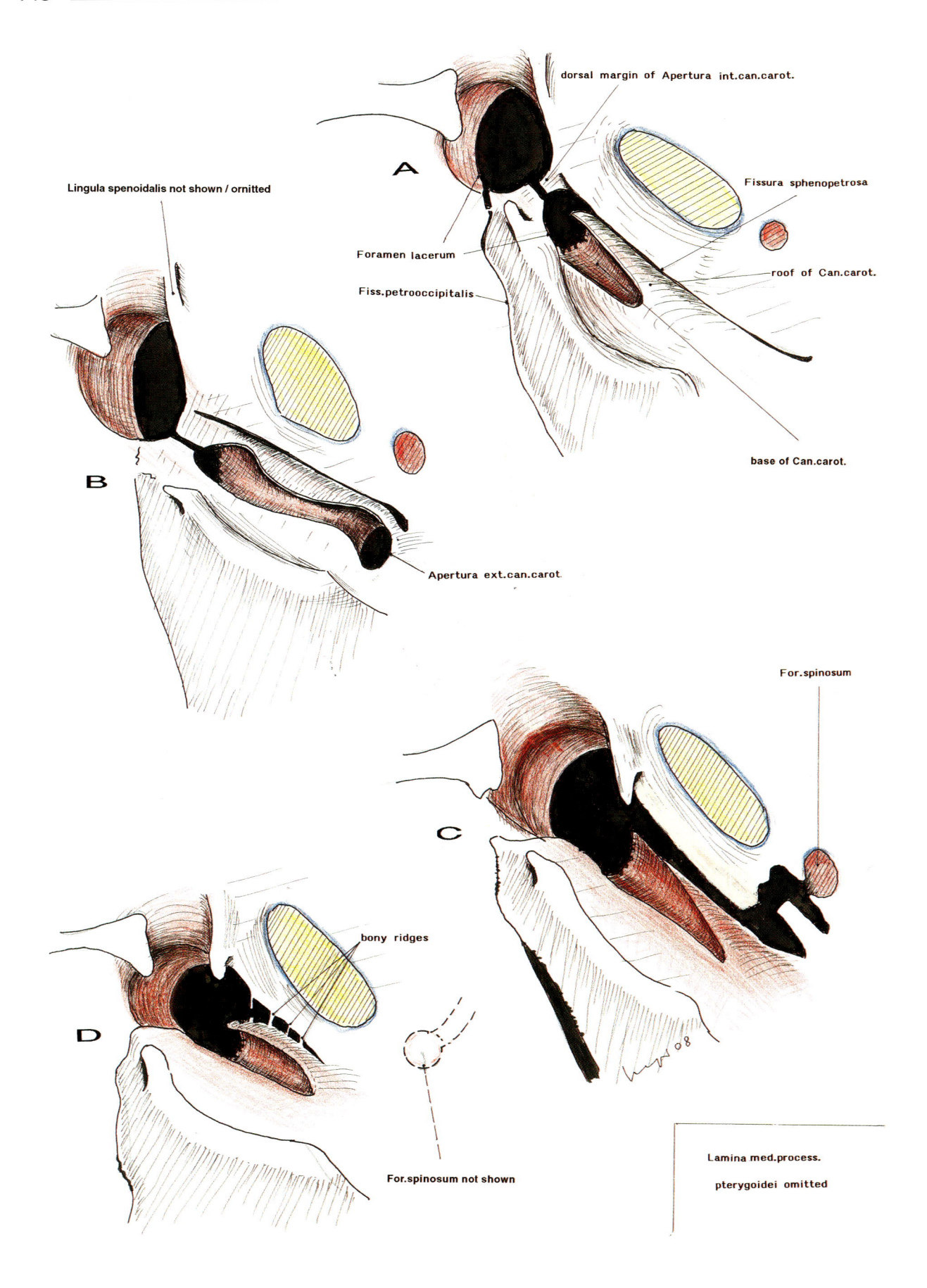

A

dorsal margin of Apertura int.can.carot.

Fissura sphenopetrosa

Foramen lacerum

roof of Can.carot.

Fiss.petrooccipitalis

base of Can.carot.

Lingula spenoidalis not shown / ornitted

B

Apertura ext.can.carot.

For.spinosum

C

bony ridges

D

For.spinosum not shown

Lamina med.process.

pterygoidei omitted

CONTENTS OF PYRAMIS, TRANSECTIONAL PLANES
(Figs. 60 to 63)

Fig. 60

Overview

A Transparent presentation according to the historic metallic cast B
B Copy of the cast of Siebenmann, presented in Rauber-Kopsch (1908) vol. 6, p 1017

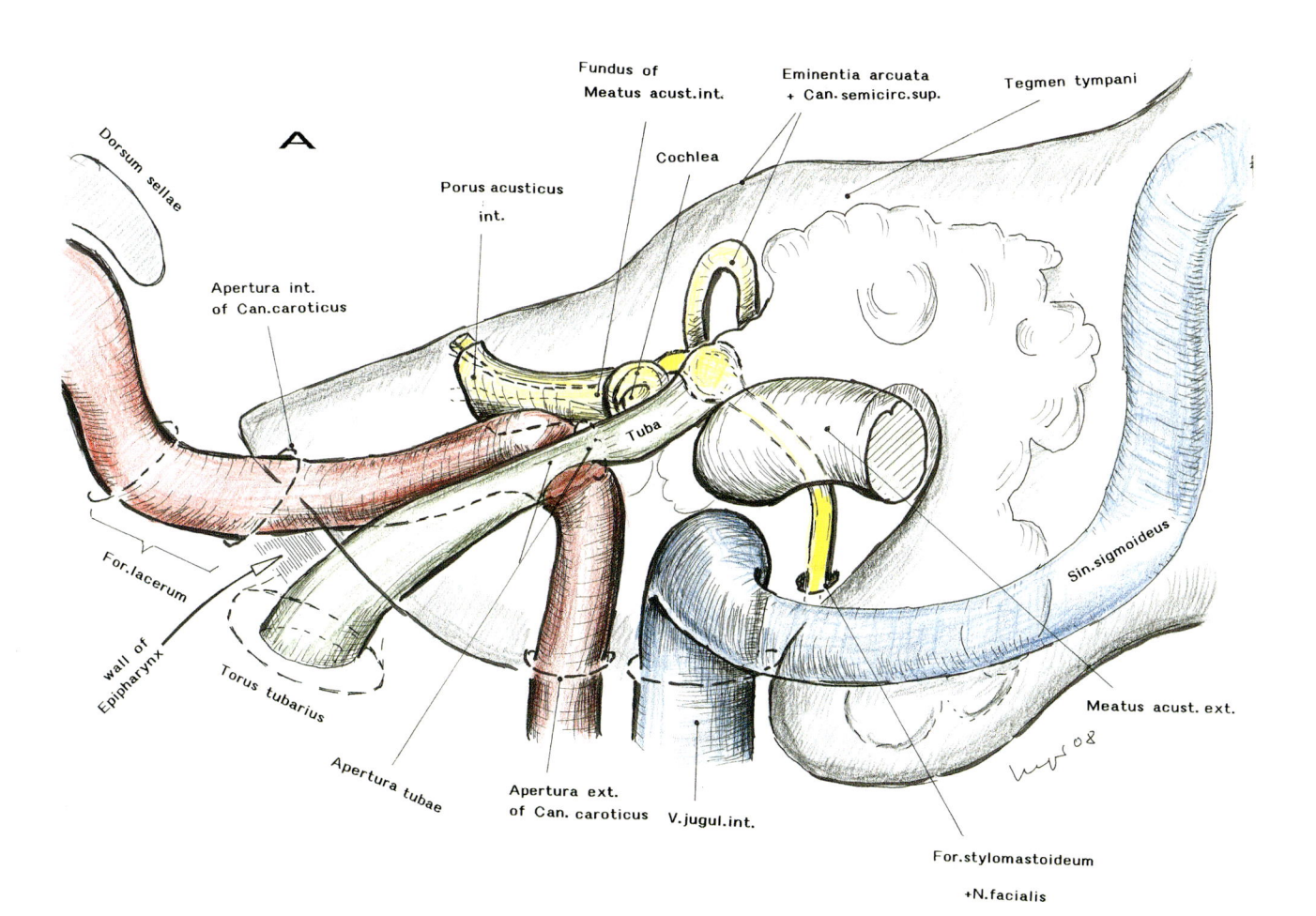

A

Dorsum sellae

Apertura int.
of Can.caroticus

Porus acusticus
int.

Fundus of
Meatus acust.int.

Cochlea

Eminentia arcuata
+ Can. semicirc.sup.

Tegmen tympani

For.lacerum

wall of
Epipharynx

Torus tubarius

Apertura tubae

Apertura ext.
of Can. caroticus

V.jugul.int.

Tuba

Sin.sigmoideus

Meatus acust. ext.

For.stylomastoideum

+N.facialis

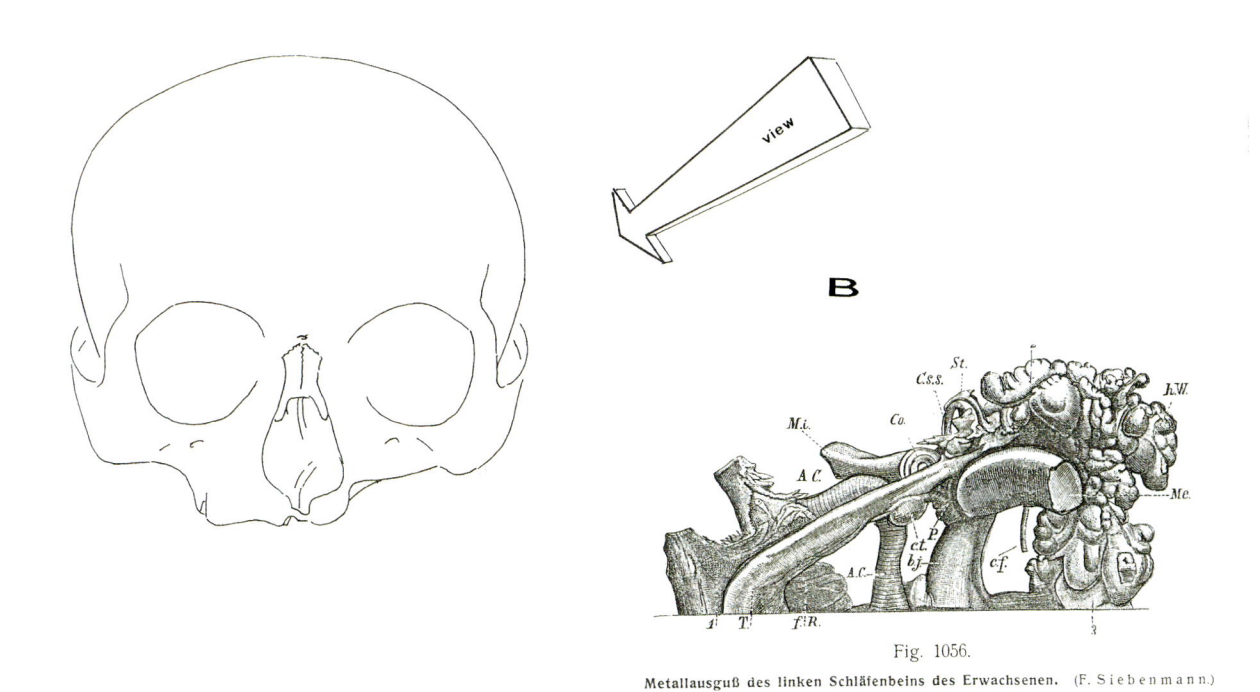

view

B

Fig. 1056.

Metallausguß des linken Schläfenbeins des Erwachsenen.　(F. Siebenmann.)

Fig. 61

A Vertical transection of the petrous bone, parallel to the upper margin, along the level of Canalis n. facialis (Fallopii).
According to a historical cadaver skull dissection (Spalteholz, 1907, p 12), modified.

B Horizontal transection of the petrous bone at the axis of Meatus acusticus externus. The horizontal segment of Canalis caroticus is known, as well as the upper area of Fossa jugularis, and its contents. According to a historical cadaver head dissection (Spalteholz, 1906, p 803), modified.

Abbreviations
1 probe in Apertura inf. canaliculi tympanici
2 distal end of probe
3 canal of N. petrosus minor (location of Jakobson's anastomosis between N.IX-N.tympanicus-N.petrosus minor)
4 Semicanalis m. tensoris tympani, lateral wall
5 begin of Semicanalis m. tensoris tympani
6 Promontorium and Tuba Eustachii

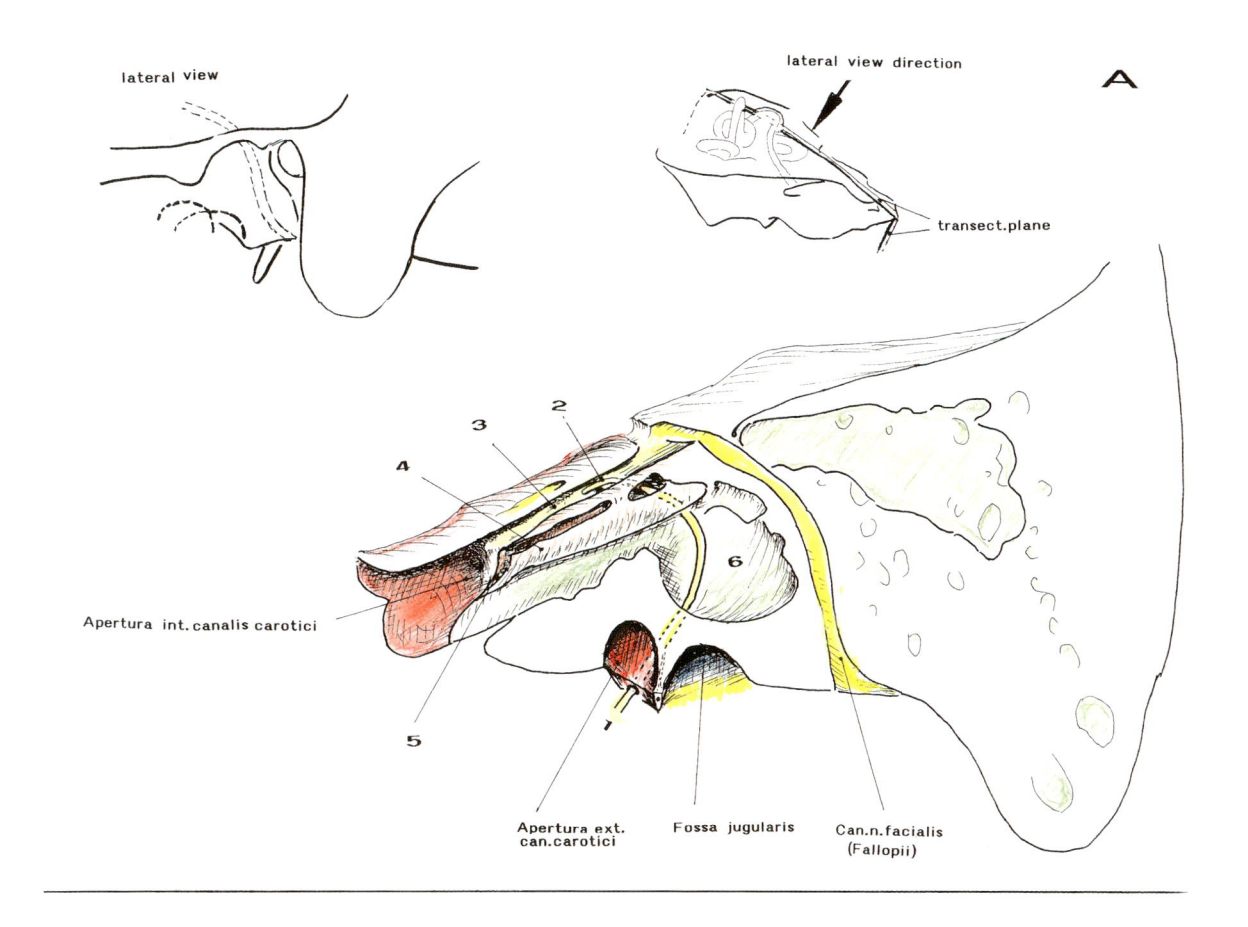

lateral view

lateral view direction

transect.plane

A

3 2

4

6

Apertura int. canalis carotici

5

Apertura ext. can.carotici

Fossa jugularis

Can.n.facialis (Fallopii)

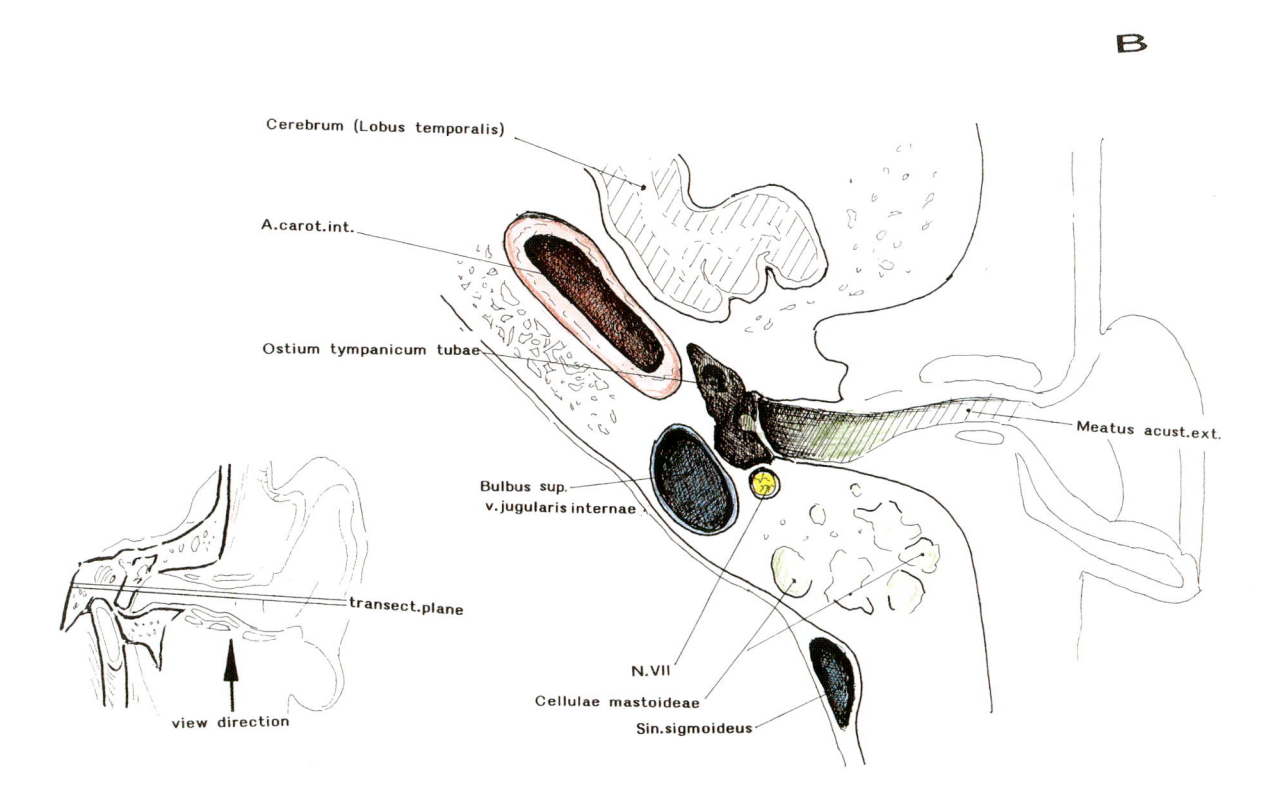

B

Cerebrum (Lobus temporalis)

A.carot.int.

Ostium tympanicum tubae

Bulbus sup. v.jugularis internae

Meatus acust.ext.

transect.plane

view direction

N.VII

Cellulae mastoideae

Sin.sigmoideus

Fig. 62

Vertical transection of Pyramis perpendicular to its longitudinal axis, at the level of Modiolus, Meatus acusticus int., and the bending segment of Canalis caroticus. Note the proximity to the base of Cochlea and the Fundus of Meatus acusticus int. The fallopian channel is located posterior to this transectional plane.

According to a cadaver skull dissection of Spalteholz (1906, p 822), modified.

Abbreviations

1   Area cochleae of Fundus meati acustici interni
2   Canalis spiralis modioli
3   Modiolus
4   Canalis longitudinalis modioli
5   Sulcus n. petrosi majoris
6   Sulcus n. petrosi minoris
7   Lamina spiralis secundaria

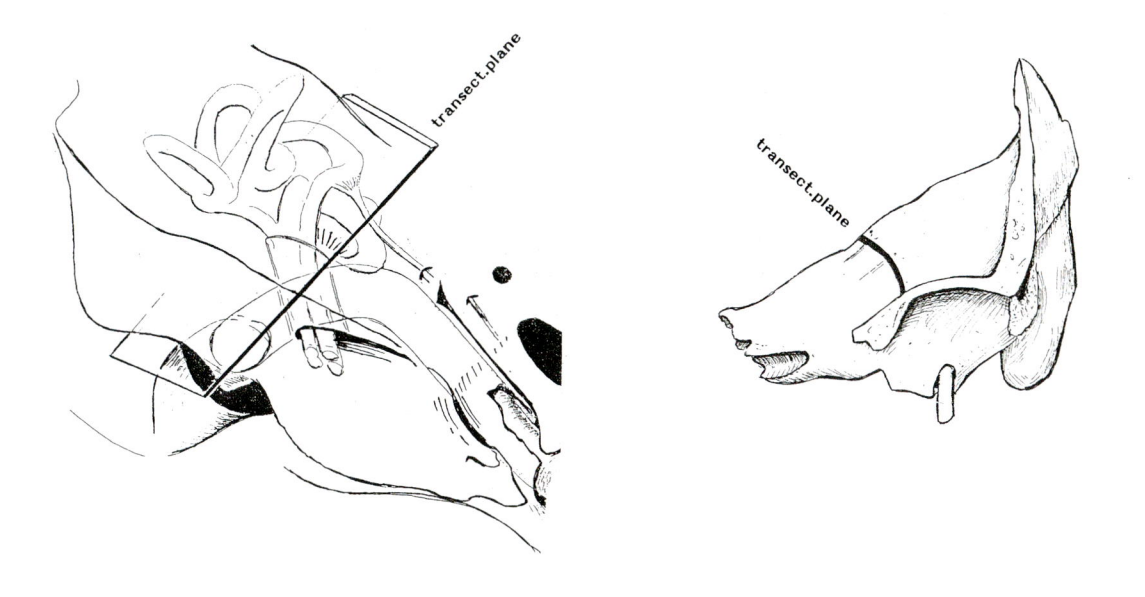

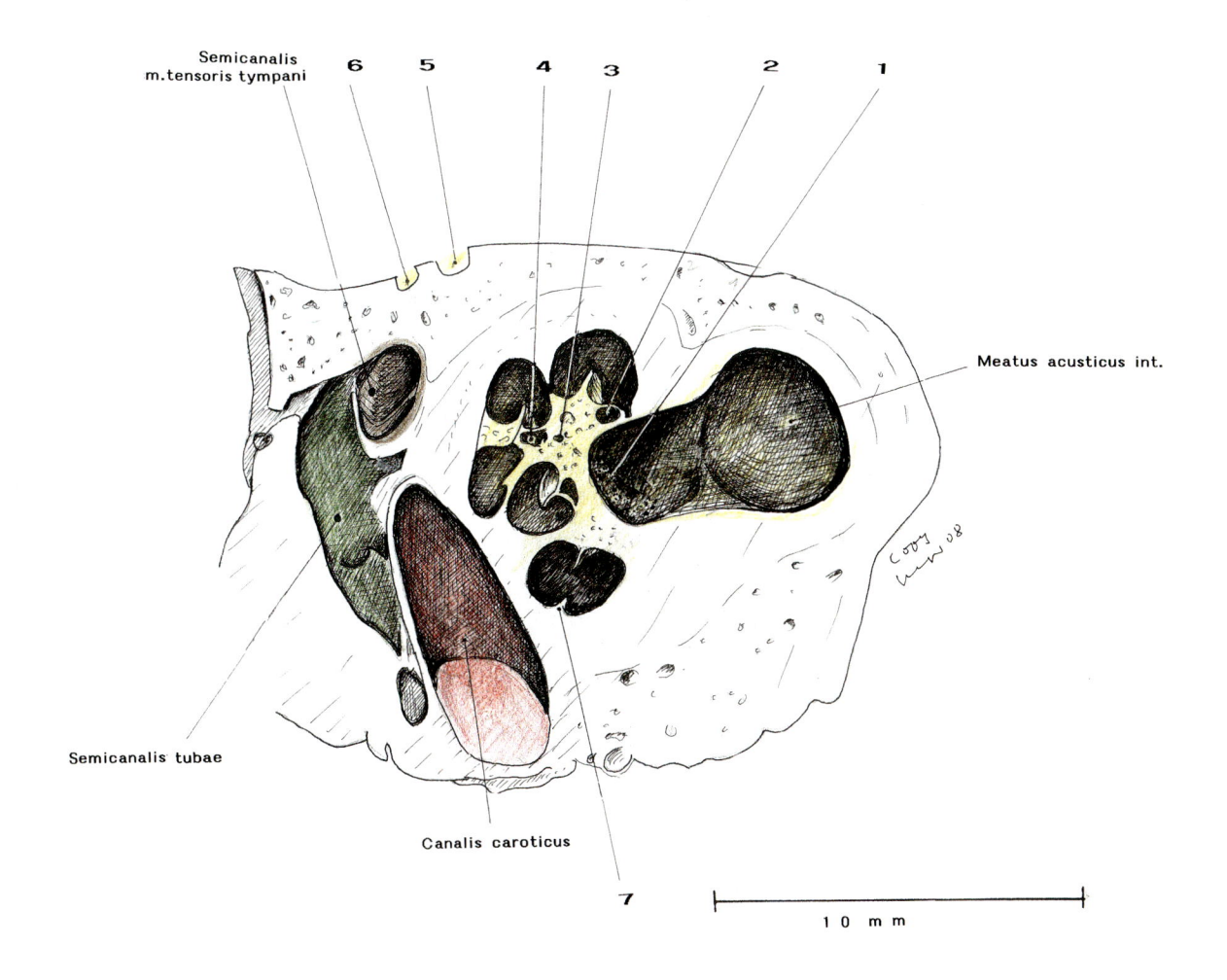

Semicanalis
m.tensoris tympani

6   **5**   **4**   **3**   **2**   **1**

Meatus acusticus int.

Semicanalis tubae

Canalis caroticus

7

10 m m

Fig. 63

Horizontal transectional plane parallel to the illustration of B in Fig. 51, at the level of Modiolus, of Canalis semicircularis lateralis, of the upper segment of Meatus acusticus ext., and the axis of Meatus acusticus int.
M. tensor tympani is transected longitudinally and dorsally, Tuba is transected longitudinally and basally.
According to a cadaver skull dissection of Spalteholz (1906, p 819), modified.

Abbreviations
1 Apertura tympanica canaliculi cochleae
2 Canalis n. facialis (Fallopii)
3 Canalis semicircularis lateralis
 a) Canalis semicircularis sup.
 b) Eminentia arcuata
4 Ampulla lat.
5 Ampulla post.
6 Area vestibularis sup.
7 Crista transversa
8 Canalis spiralis modioli
9 Lamina spiralis ossea
10 Modiolus
11 Lamina modioli
12 Semicanalis m. tensoris tympani
13 Semicanalis tubae and its Cellula pneumatica
14 Promontorium
15 Prominentia styloidea
16 bony wall between Tympanon and Foramen jugulare
17 Sulcus tympanicus

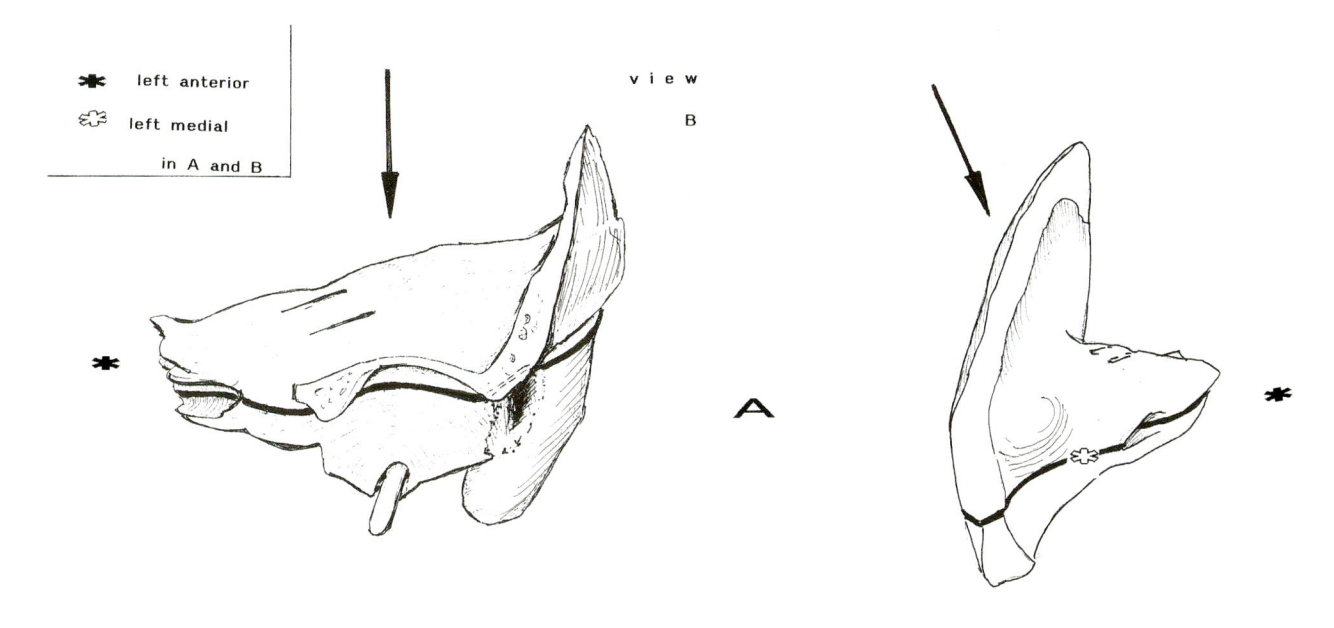

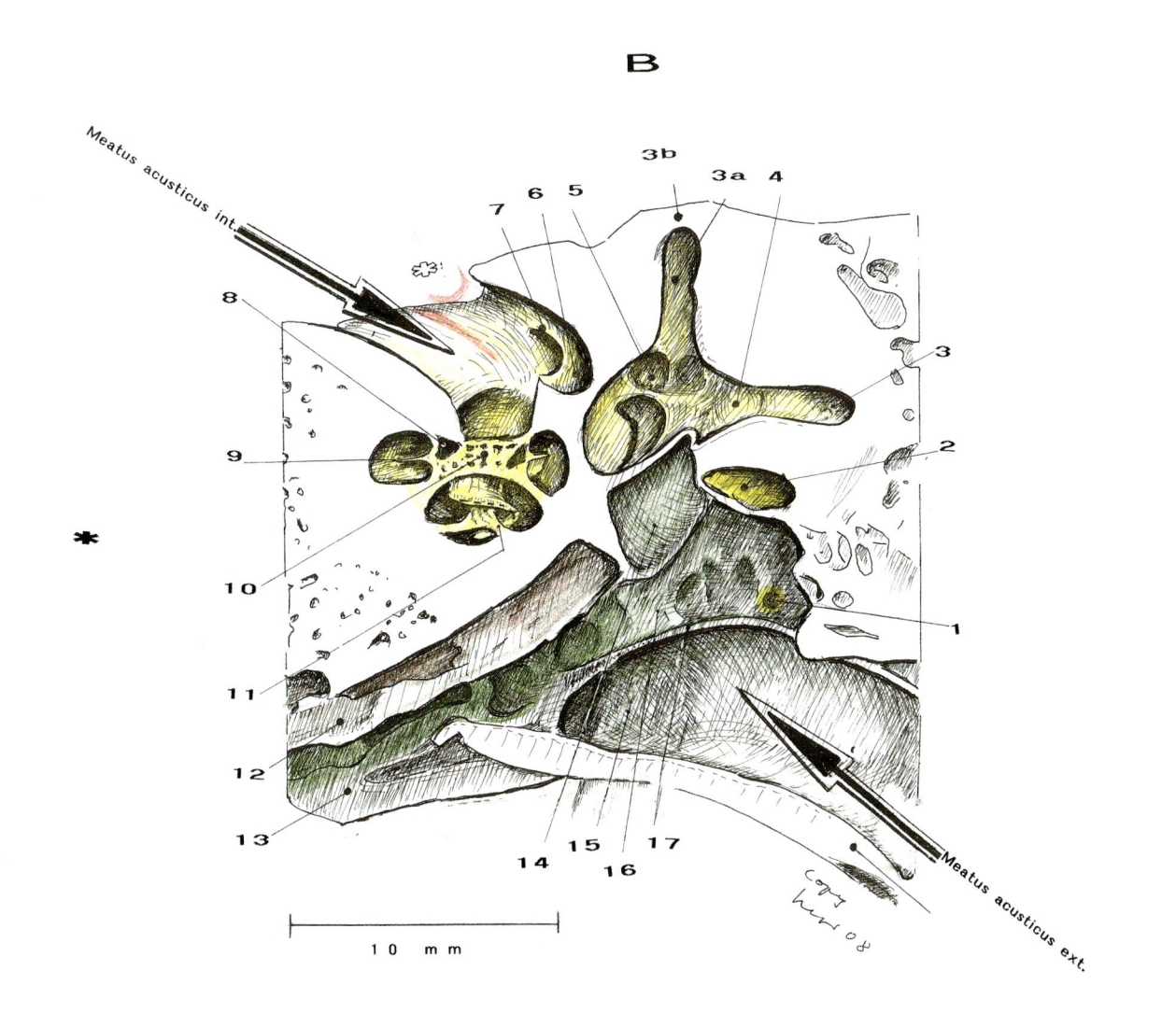

left anterior
left medial
in A and B

view
B

A

B

Meatus acusticus int.

Meatus acusticus ext.

10 mm

Literature

Caix MN, Outrequin G (1979) Variability of the bony semicircular canals. Anat Clin 1, 259–265

Cavallo LM, Messina A, Cappabianca P, Esposito F, de Divitis E, Gardner P, Tschabitscher M (2005) Endoscopic endonasal surgery of the midline skull base. Anatomical study and clinical considerations. Neurosurg Focus 19 (1), E2, pp 1–14

Donaldson I (1980) Surgical anatomy of the tympanic nerve. J Laryngol Otol 94, 163–168

Evans W (1956) Carotid canal anomaly. Other instances of absent internal carotid artery. Med Times 84/10, 1069–1072

Goodman RS, Cohen NL (1981) Aberrant internal carotid artery in the middle ear. Ann Otol Rhinol Laryngol 90, 67–69

Helms J (1978) Varianten der Gefäße im Meatus acusticus int. In: Plester D, Wende S, Nakayama N eds. Springer, Berlin Heidelberg New York

Helms J (1981) Variations of the course of the facial nerve in the middle ear and mastoid. In: Samii M, Janetta FJ eds. The cranial nerves. Springer, Berlin Heidelberg New York

James TM, Presley R, Steed F (1980) The foramen ovale and sphenoidal angle in man. Anat Embryol (Berl) 160, 93–104

Kassam A, Snyderman CH, Mintz A, Gardner P, Carrau RL (2005) Expanded endonasal approach: The rostrocaudal axis. Part II. Posterior clinoides to the foramen magnum. Neurosurg Focus Jul 15; 19(1): E4

Kassam AB, Gardner P, Snyderman C, Mintz A, Carrau R (2005) Expanded endonasal approach: Fully endoscopic, completely transnasal approach to the middle third of the clivus, petrous bone, middle cranial fossa, and infratemporal fossa. Neurosurg Focus Juo; 15:19 (1):E6

Krmpotic-Nemanic J, Draf W, Helms J (1985) Chirurgische Anatomie des Kopf-Hals-Bereichs. P 225 pyramis, pp 222–223 arteries of labyrinth. Springer, Berlin Heidelberg New York Tokyo

Krmpotic J, Nikolic V (1964) Praktisch wichtige topographische Beziehungen des N. petrosus superficialis major in der mittleren Schädelgrube. Z Laryngol Rhinol Otol 43/12, 748–753

Portmann M, Sterkers JM, Charachon R, Chouard CH (1975) The interne auditory meatus. Anatomy, pathology and surgery. Livingstone, Edinburgh London New York

Proctor B, Nager ST (1982). The facial canal: Normal anatomy, variations and anomalies. I. Normal anatomy of the facial canal. Ann. Otol. 1, 44

Rauber-Kopsch (1908) Bd. 6, p 1017. Thieme, Leipzig

Schenk D, Seeger W (1968) Otologische Komplikationen nach Kirschner'scher Elektrokoagulation des Ganglion Gasseri. Tagung Dtsch. Ges. Hals-Nasen-Ohrenheilk. Bad Reichenhall 27.5.

Siebenmann see Rauber-Kopsch (1908)

Spalteholz W (1906) Handatlas der Anatomie des Menschen, pp 502-509, p 803, p 810, p 819, p 822. Hirzel, Leipzig

Thomas JR (1980) Tympanic neurectomy and chorda tympani section. Aust NZJ Surg 50, 352–355

CHAPTER VI
CLIVUS AREA AND PARS CONDYLARIS
(Figs. 64 to 70)

Overview (Figs. 64 and 65)

In clinical terminology, Clivus means Clivus Blumenbachii and its underlying bony segment between the posterior limit of the roof of Choana and the anterior margin of Foramen occipitale and the bony bloc of the Foramen jugulare-condylar-complex

Components of the Clivus area

Clivus Blumenbachii is the intracranial medial plane between Dorsum sellae and the anterior margin of Foramen occipitale (anatomical nomenclatura).

The clinical defining includes 3 segments:

- Its small rostral segment is the caudal part of **Corpus sphenoidale**. Corpus sphenoidale is the central part of Os sphenoidale enclosing Sinus sphenoidalis. It is connected to the large caudal segments by Synchondrosis sphenooccipitalis in children and adolescents (Fig. 3), and it merges with
- **Pars basilaris** of the occipital bone in adults.

  Merging of Corpus sphenoidale to adjacent components of the sphenoid bone occurs at the early years of life, earlier than merging of Corpus sphenoidale and Pars basilaris. The caudal-lateral component of the so-called Clivus is **Pars condylaris**. It connects to Pars basilaris and to Squama occipitalis (tabular portion, Gray, ed. 1974) at the same period as the components of the sphenoid bone, earlier than Corpus sphenoidale and Pars basilaris of the occipital bone.

Phylogenetic, ontogenetic and dysontogenetic aspects (Fig. 66)

Lesions of the Clivus area are today more relevant than in the past, especially in endoscopic transnasal surgery. Dysontogenetic tumors, Clivus chondromas and Clivus chordomas are the most important entities. These tumors originate from phylo- and ontogenetic residuals, which form Synchondrosis sphenooccipitalis. This is a chondroid layer, which covers a component of Chorda dorsalis, similar to intervertebral discs.

Synchondrosis in children and adolescents also plays a role in neuronavigation (Fig. 66).

Basal extracranial Clivus area (Figs. 66 to 68)

Fissura petrooccipitalis is the lateral border of Pars basilaris of the occipital bone. This wide deep bony gap is filled and flattened by Enchondrosis petrooccipitalis, as demonstrated in Fig. 38.

The lateral border of the condylar segment of the occipital bone is formed by Foramen jugulare and its Fossa (medial shapes). The lateral boundary is Pyramis. The extension of Fissura petrooccipitalis is a fine suture at the base of Fossa jugularis, which is the starting point of Sutura occipitomastoidea. Bony dysplasias may be relevant for surgery.

Bony structures forming
- Pars condylaris (bilateral)
- Pars basilaris
- Pars squamosa (tabular part, Gray ed. 1974)

These components are similar to the Atlas-Dens-complex in newborns.
Atlas and occiput may merge partially or completely. An occipital vertebra may develop from components of the occipital bone, partially or completely. These variants are well known in neurosurgery because these variants may result in brainstem compressions.

These variants help to understand the variability of Foramen jugulare as presented by Helms (1978). Some examples of cadaver skull dissections are shown in Fig.68.

Dorsal intracranial Clivus area (= Clivus Blumenbachii plus condylar part) (Figs. 4, 5, and 65)

In its intracranial contour, Fissura petrooccipitalis is flat and small in contrast to the extracranial shape. But there the flat Sulcus petrosus inf. is located (Fig. 77). Sinus petrosus inf. is wider than its bony sulcus. Before entering Bulbus sup. v. jugularis it divides into branches, especially at the medial area of the anterior segment of Foramen jugulare. The cranial nerves IX to XI are located medially to it. V. emissaria condylaris connects the venous branches of Fossa paracondylaris to Bulbus sup. v. jugularis internae. Cranial nerves are separated from the surrounding veins by arachnoid tunnels.
These are extensions of cisternal CSF spaces similar to CSF spaces around other craniospinal nerves.
The Tuberculum jugulare-condylar-bloc is located posterior-laterally to the Clivus. It encloses Canalis n. hypoglossi and its numerous venous branches and venous emissaries (Figs. 69 and 70).

CLIVUS AREA AND PARS CONDYLARIS (Figs. 64 to 70)

Fig. 64

Basal bony bloc below Clivus
– segment of Corpus sphenoidale posterior to the roof of Choana
– Pars basilaris of Os occipitale below Clivus

Pars condylaris of Os occipitale

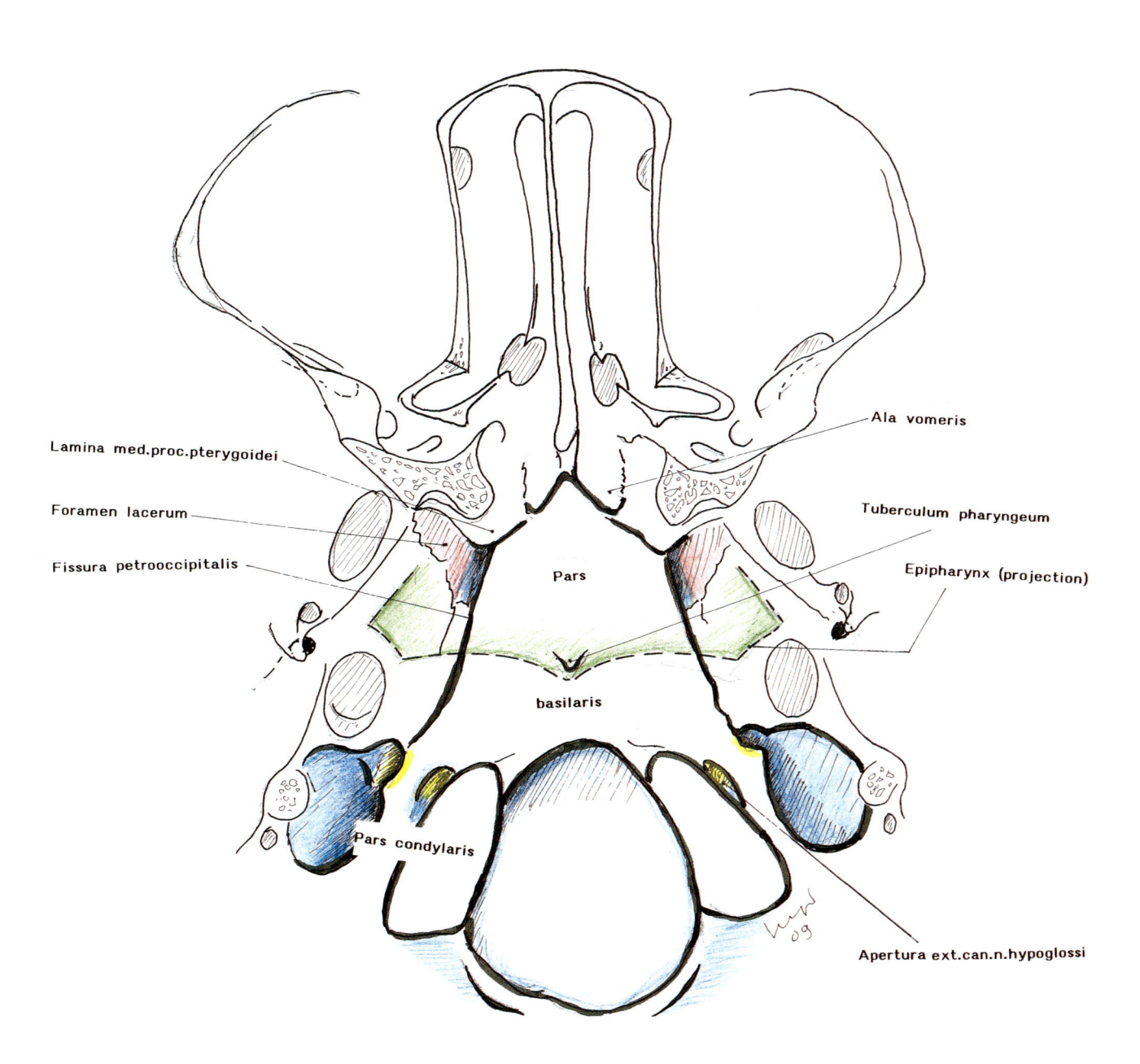

Lamina med.proc.pterygoidei

Foramen lacerum

Fissura petrooccipitalis

Ala vomeris

Tuberculum pharyngeum

Epipharynx (projection)

Pars

basilaris

Pars condylaris

Apertura ext.can.n.hypoglossi

✳ For.sphenopalatinum

Fig. 65

Clivus Blumenbachii and Pars condylaris
–   segment of Corpus sphenoidale (postsellar segment)
–   Pars basilaris of Os occipitale

Pars condylaris of Os occipitale
–   Tuberculum jugulare-condylar-bloc

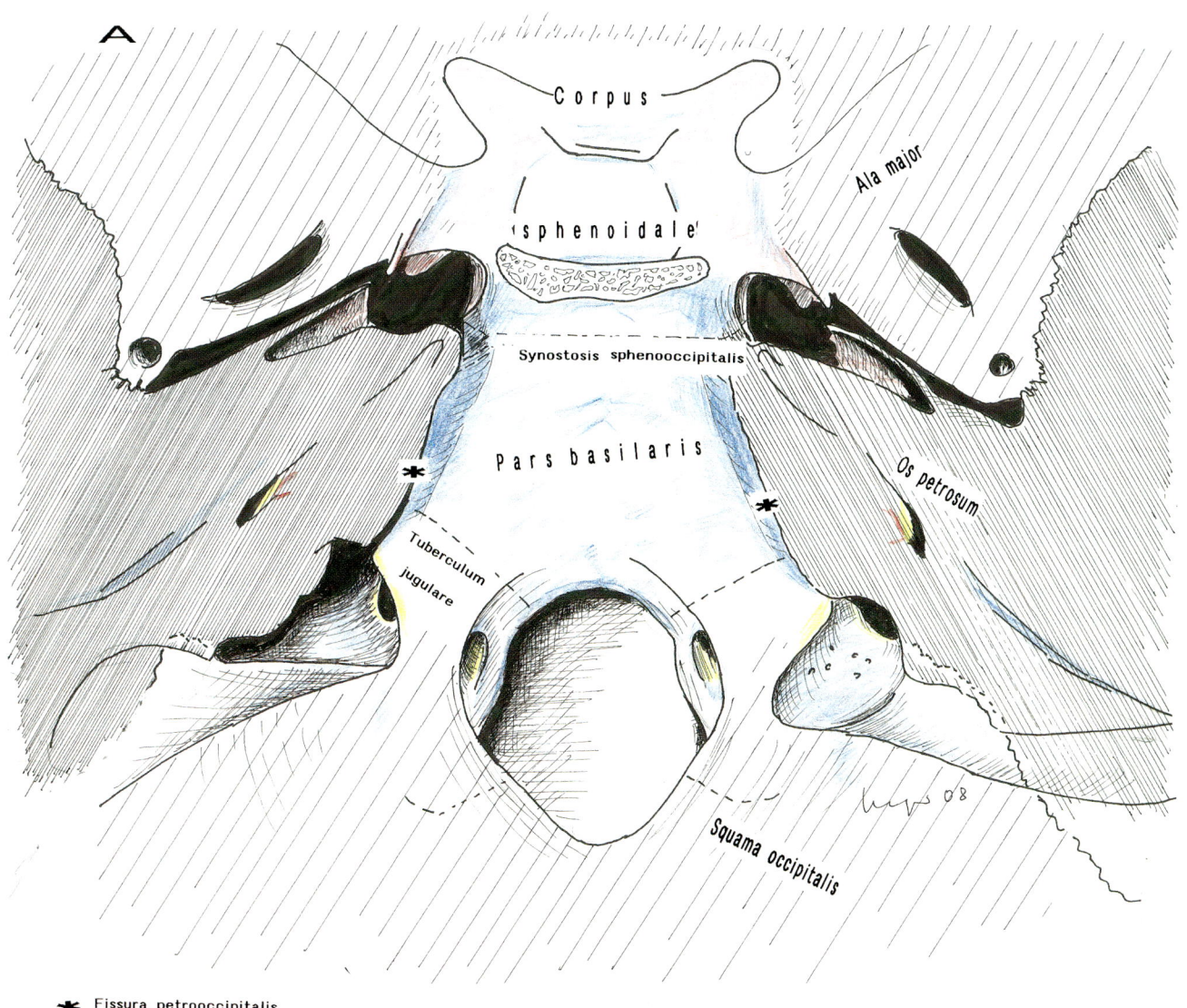

Fissura petrooccipitalis

Fig. 66

Skull base of a child (12 years)

A   Synchondrosis can be used as a neuronavigatory landmark
B   Cranial base, sphenoidal and occipital segments of the bony bloc forming Clivus

Abbreviations
1   bony gap, usual finding before puberty
2   Corpus sphenoidale
3   Pars basilaris of the occipital bone
4   and 4' Fissura sphenopetrosa a wide; usual finding in children and often in adults
5   Concha inferior
6   Foramen sphenopalatinum
7   Fissura petrooccipitalis
8   Apertura externa canalis carotici
9   Apertura tubae (projection)

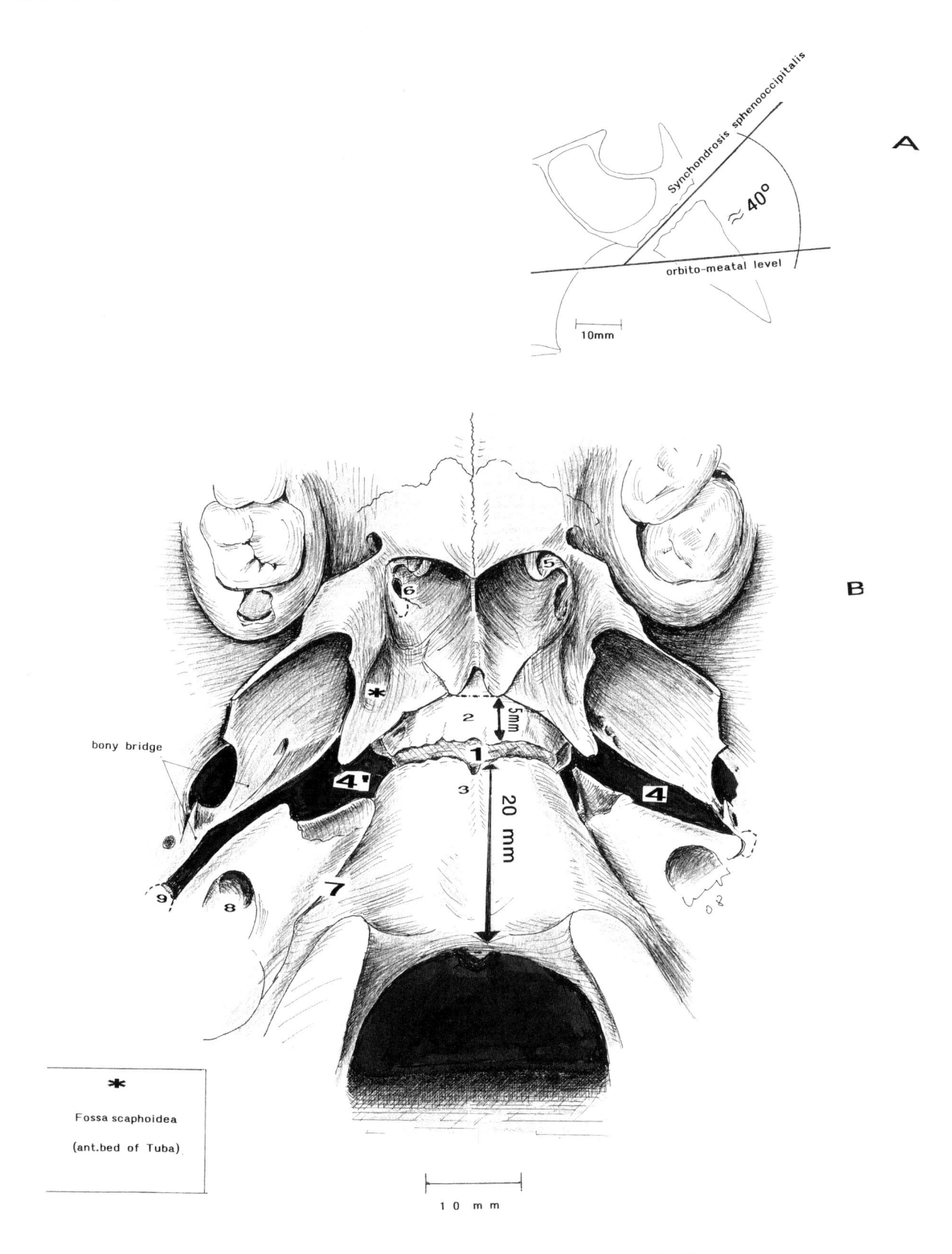

Synchondrosis sphenooccipitalis

≈40°

orbito-meatal level

10mm

A

B

bony bridge

*

2

5mm

1

4'

3

20 mm

4

5

6

7

8

9

08

*

Fossa scaphoidea

(ant.bed of Tuba)

10 mm

gap for Synchondrosis sphenooccipitalis in
infants and juveniles — 1 —

Fig. 67

Foramen jugulare and surrounding structures

A Overview
B Sectional enlargement

Abbreviations
1 Foramen ovale
2 Foramen spinosum
3 Sutura petrosquamosa
4 Fissura petrotympanica
5 Apertura externa canalis carotici
6 Porus acusticus externus
7 Processus styloideus cut
8 Foramen stylomastoideum
9 Fossa jugularis
10 Processus mastoideus
11 Sutura occipitomastoidea
12 Fossa paracondyloidea
13 Condylus occipitalis
14 Foramen occipitale
15 Apertura externa canalis nervi hypoglossi
16 Fissura petrooccipitalis
17 Apertura tubae
18 Fissura sphenopetrosa
19 Processus vaginalis
20 as 11
21 Processus paracondyloideus
22 Processus intrajugularis
23 connection of Processus paracondyloideus to Condylus occipitalis
24 posterior segment of Foramen jugulare (for V. jugularis interna)
25 anterior segment of Foramen jugulare (for cranial nerves-medially-, and for ramifications of Sinus petrosus inf.-laterally-)

FIG. 67

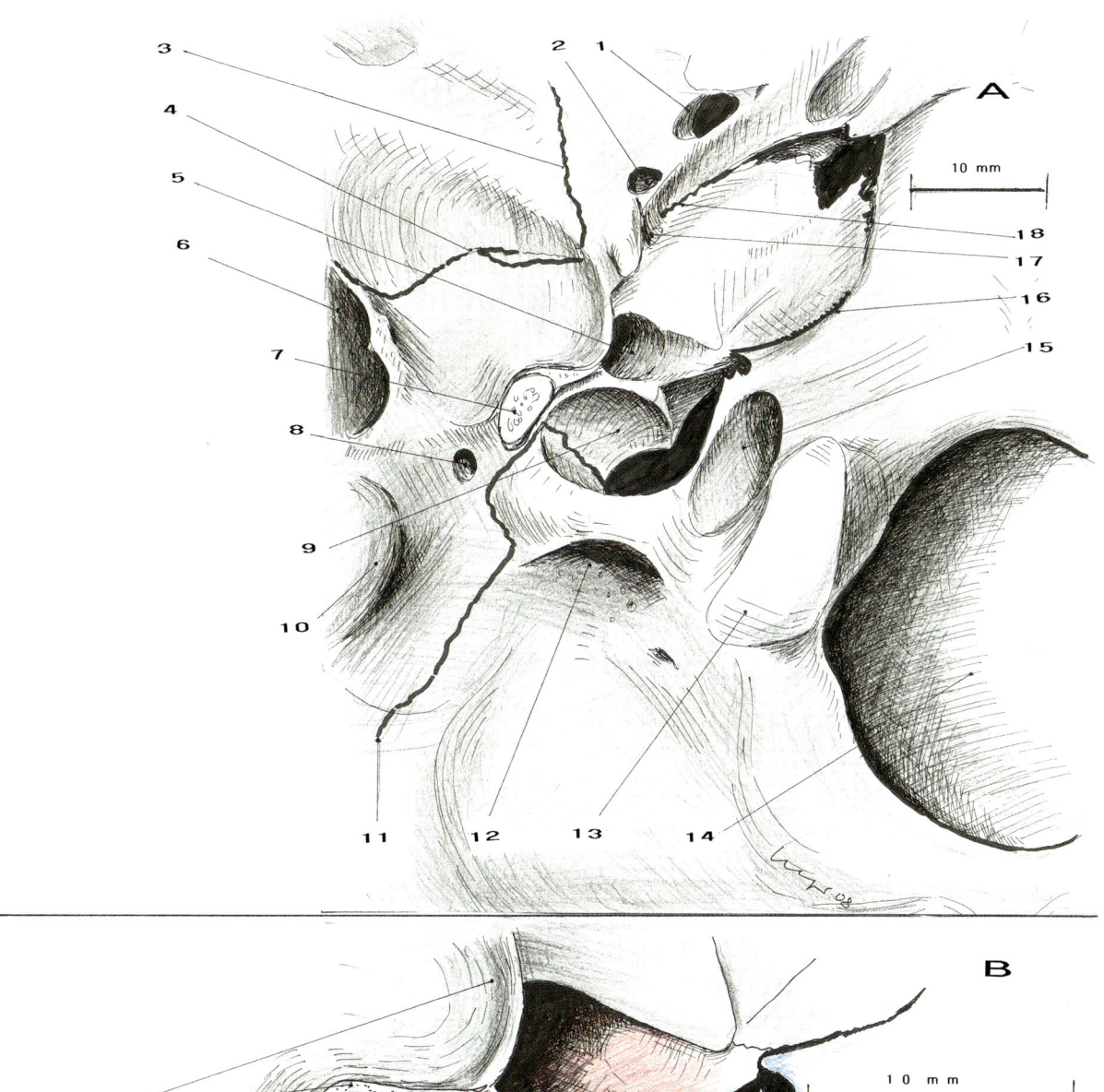

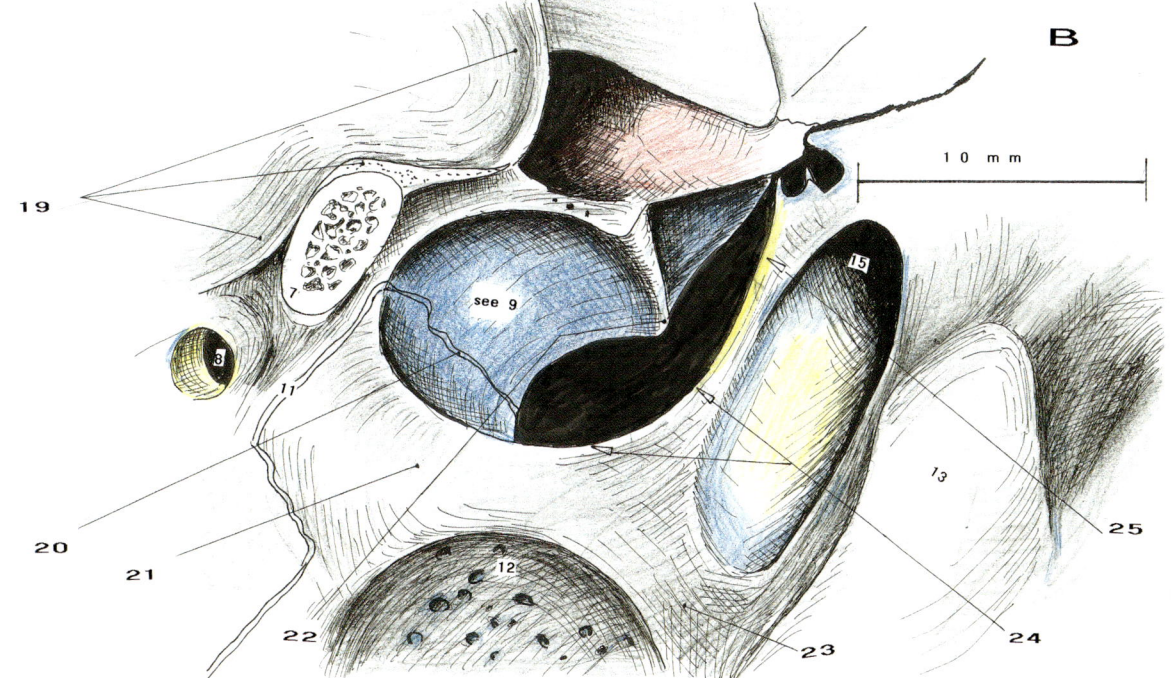

Fig. 68

Addendum to Fig. 67. – Examples

A 20 y, female. Tuberculosis. Chronic high venous pressure.
B 48 y, female. Close relationship of Processus styloideus to Foramen jugulare (usual finding)
C 65 y, male. Senile atrophy. Probes in Canales n. hypoglossi. Bony defect between Fossa jugularis and Apertura ext. n. hypoglossi on the left side (see probe).

Abbreviations
1   Fossa paracondyloidea
2   Fossa jugularis
3   Foramen stylomastoideum
4   Processus styloideus
5   Processus vaginalis
6   Apertura ext. canalis carotici
7   Spina angularis
8   Foramen spinosum
9   Apertura ext. canalis n. hypoglossi
10 Tuberculum pharyngeum
11 Fissura (Synchondrosis) petrooccipitalis
12 Apertura of Pars ossea tubae
13 Fissura (Synchondrosis) sphenopetrosa
14 roof of Apertura ext. canalis carotici
15 Processus intrajugularis
16 Foramen jugulare, posterior segment, enclosing Bulbus sup. v. jugul. internae
17 Foramen jugulare, anterior segment, enclosing cranial nerves IX to XI

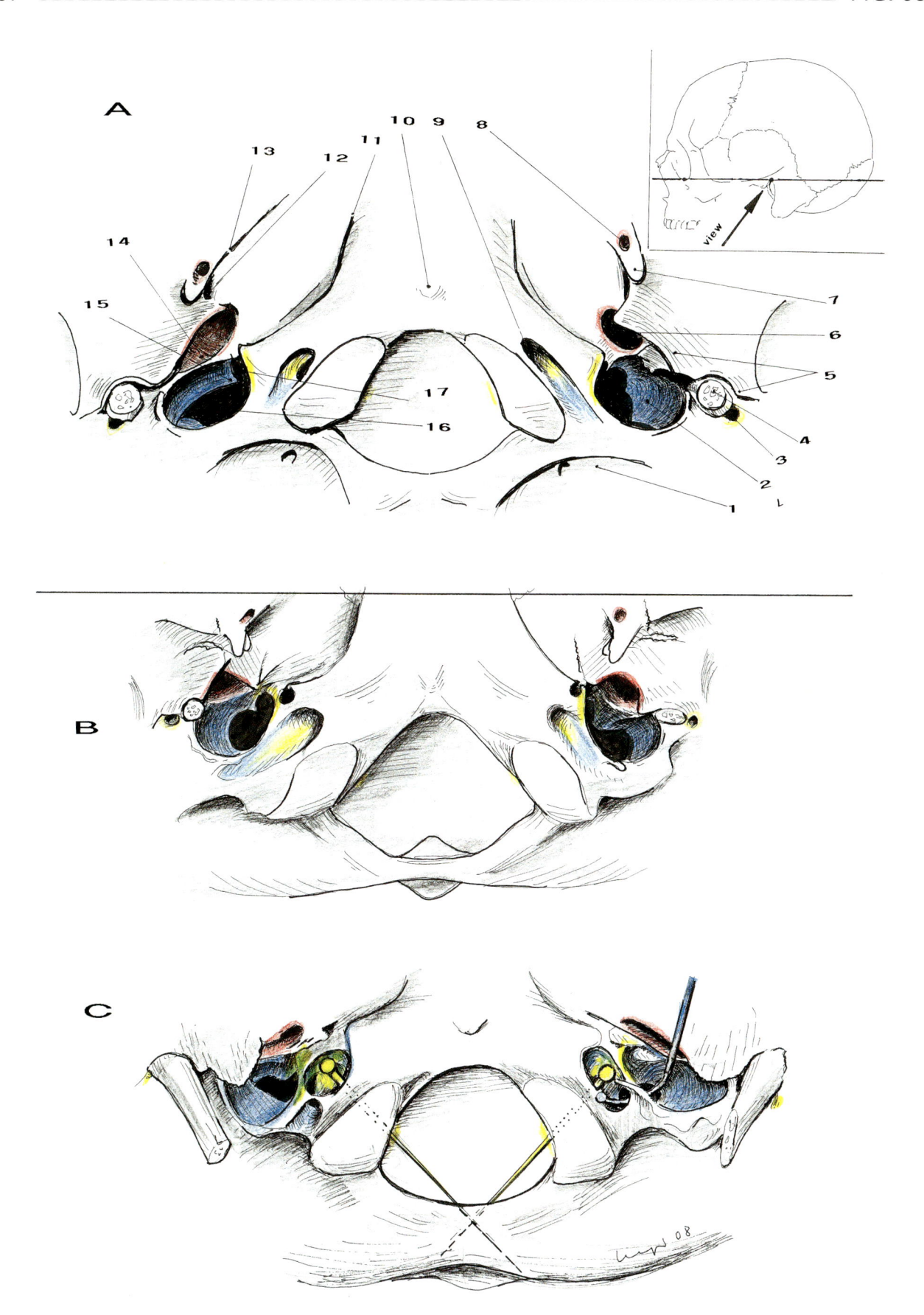

Fig. 69

Common variant

Probes: Venous channels are connecting Canalis n. hypoglossi to Foramen and Fossa jugularis)

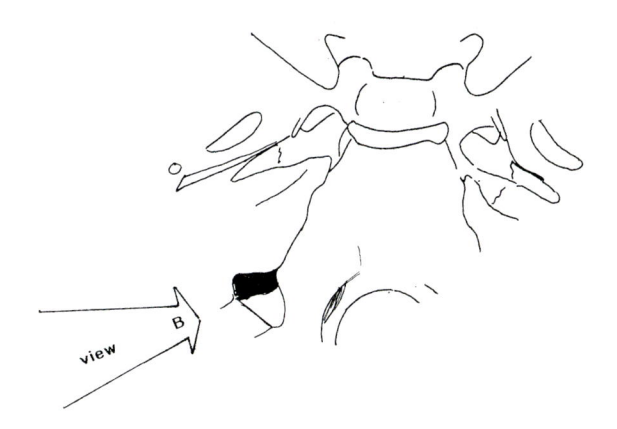

view B

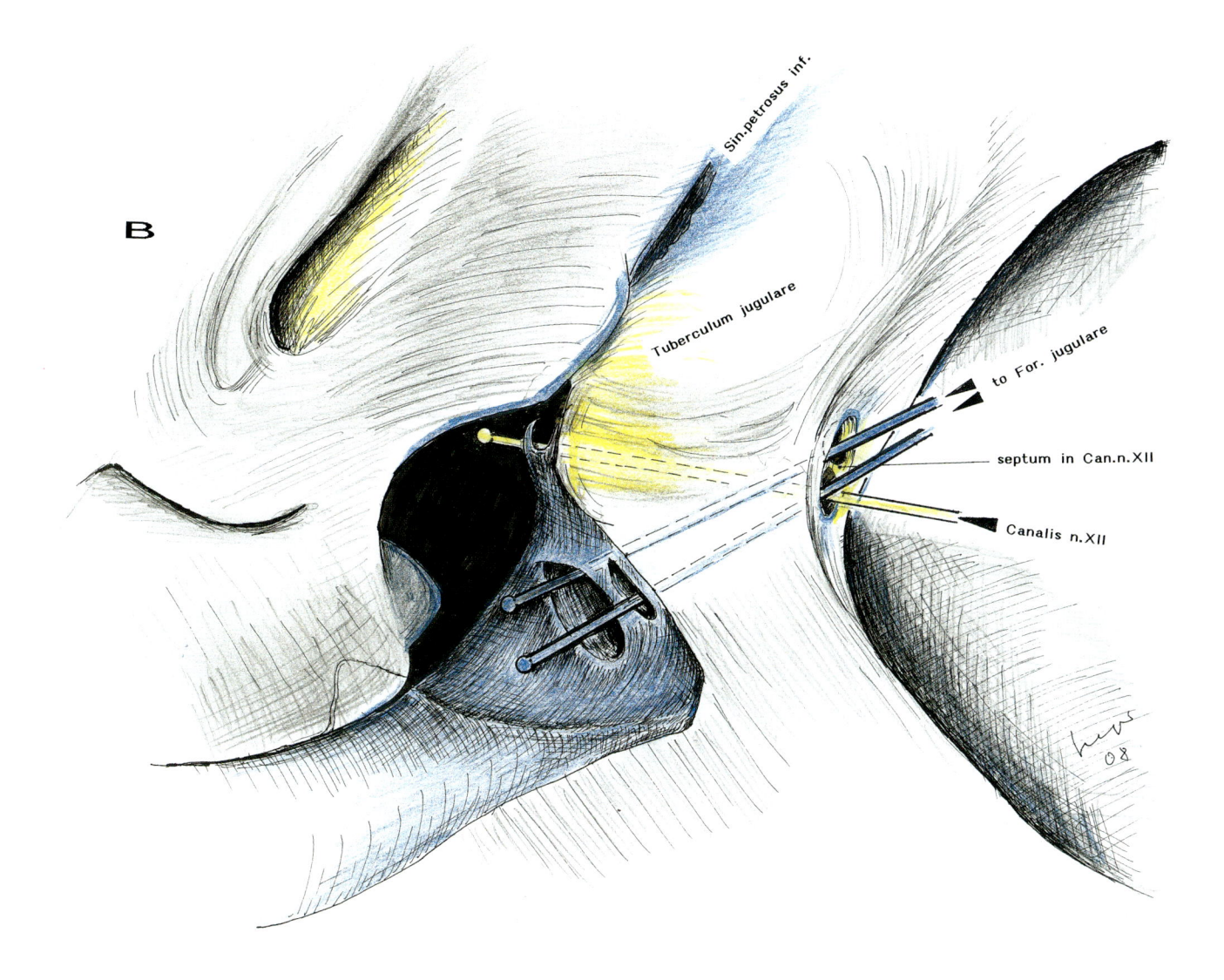

B

Sin. petrosus inf.

Tuberculum jugulare

to For. jugulare

septum in Can. n. XII

Canalis n. XII

Fig. 70

Interforaminal cadaver skull transection

At this transectional plane the following structures are located on one straight line:

-   posterior margin of Porus acusticus int.
-   anterior segment of Foramen jugulare (not visible on CT in many cases)
-   anterior margin of Apertura int. of Canalis n. hypoglossi

A Overview
B Sectional enlargement

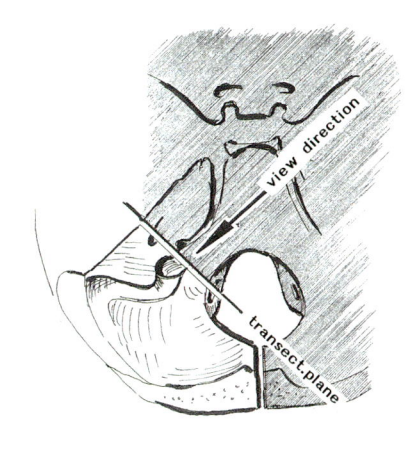

topogram

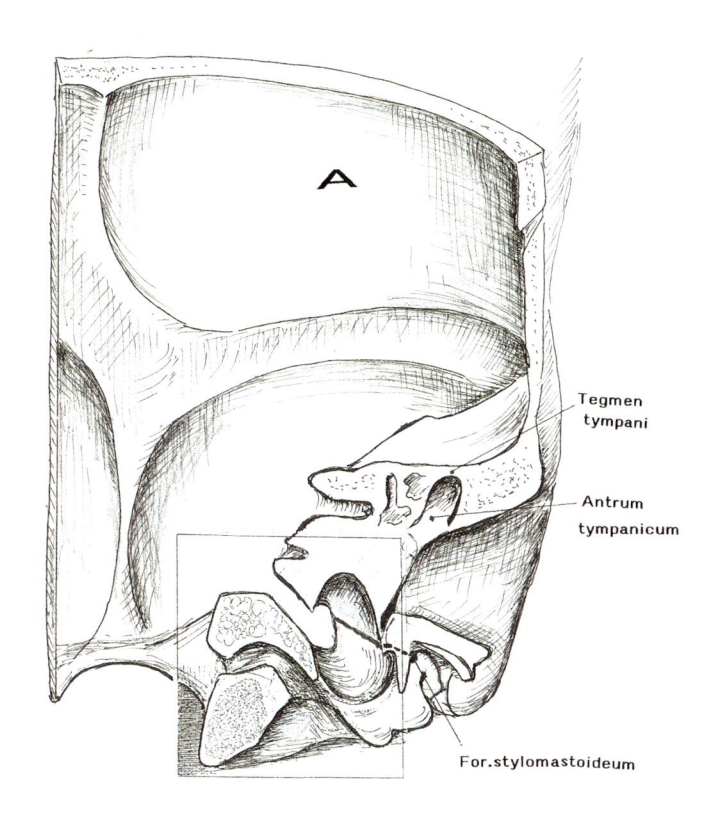

A

Tegmen
tympani

Antrum
tympanicum

For.stylomastoideum

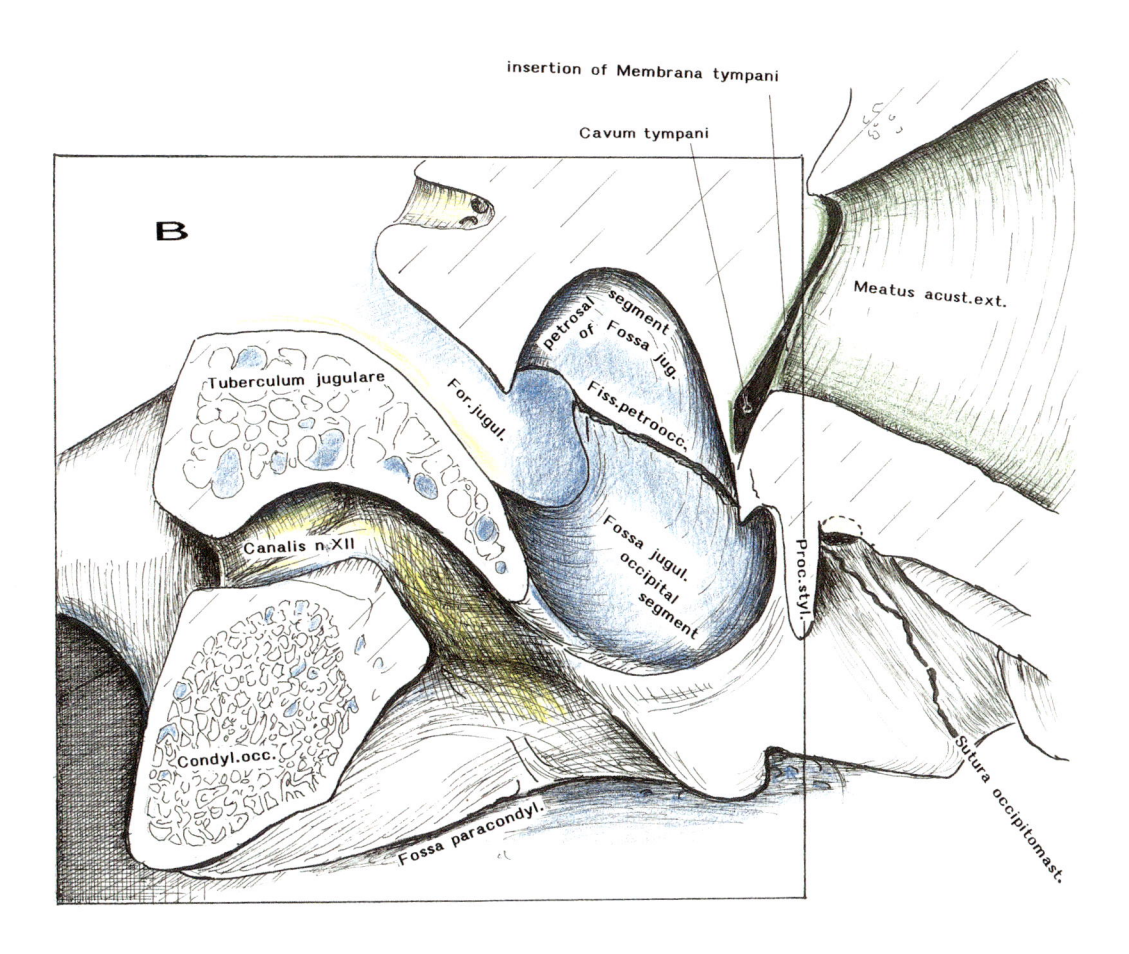

B

insertion of Membrana tympani

Cavum tympani

Meatus acust.ext.

Tuberculum jugulare

For.jugul.

petrosal segment of Fossa jug.

Fiss.petroocc.

Fossa jugul. occipital segment

Canalis n.XII

Proc.styl.

Condyl.occ.

Sutura occipitomast.

Fossa paracondyl.

Literature

Cavallo LM, Messina A, Cappabianca P, Esposito F, de Divitis E, Gardner P, Tschabitscher M (2005) Endoscopic endonasal surgery of the midline skull base. Anatomical study and clinical considerations. Neurosurg Focus 19 (1), pp 1–14

Corlieu P, Aaron C, Godefrov D (1981) Radioanatomie de la base du crâne, étage moyen et postérieur. Ann Otolaryngol Chir Cervicofac 98, 173–179

Kassam AB, Gardner P, Snyderman C, Mintz A, Carrau R (2005) Expanded endonasal approach: Fully endoscopic, completely transnasal approach to the middle third of the clivus, petrous bone, middle cranial fossa and infratemporal fossa. Neusorug Focus Jul; 15:19(1):E6

Kassam A, Snyderman CH, Mintz A, Gardner P, Carrau RL (2005) Expanded endonasal approach: The rostrocaudal axis. Part II. Posterior clinoides to the foramen magnum. Neurosurg Focus Jul 15; 19(1): E4

Lang J (1979) Kopf, Teil B, Gehirn- und Augenschädel. Springer, Berlin Heidelberg New York

Lang J (1981) Neuroanatomie der Nn opticus, trigeminus, facialis, glossopharyngeus, vagus, accessorius und hypoglossus. Arch Otorhinolaryngol 231, 1–69

Lang J, Schafhauser O, Hoffmann S (1983) Über die postnatale Entwicklung der transbasalen Schädelpforten: Canalis caroticus, Foramen jugulare, Canalis hypoglossalis, Canalis condylaris und Foramen magnum. Anat Anz 153, 315–357

Rauber-Kopsch (1906) Lehrbuch der Anatomie, Bd. 2. Thieme, Leipzig

Samii M, Janetta PJ (1981) The cranial Nerves. Springer, Berlin Heidelberg New York

Spalteholz W (1906) Handatlas der Anatomie des Menschen, Band 3, pp 504–505. Hirzel, Leipzig

CHAPTER VII
SPECIAL SURGICAL ASPECTS. EXAMPLES
(Figs. 71 to 79)

Planning strategies (Figs. 71 to 77)

Imaging techniques (Figs. 71 to 74)

A cadaver skull transection at the level of the interforaminal line connecting Porus acust. int. – For. jugulare – Apertura int. n. XII (see Fig. 70) was merged with MRT/CT-slices, for a better understanding of the asymmetric slices. These inconsistencies may occur by incorrect positioning of the head or by asymmetries of the head and skull.

Measurements (Figs. 75 to 77)

Anatomical landmarks may be unclear. Distance estimations may be useful for orientation, especially for defining the margins of Os basilare and of the condylar part.

Dural penetration point of N. abducens (Figs. 76 and 77)

A further problem is the neuronavigatory definition of the dural penetration point of N. abducens. This area is located close to some essential structures (Fig. 76)
It is crucial to identify in skull base surgery, especially of petroclival meningiomas. The tumor usually masks the nerve along its intradural-transtumoral course. Identification of the nerve is easier during an anterior than an posterior surgical approach, as in transnasal endoscopy. Before the nerve enters the lateral wall of the cavernous area of the carotid artery, it crosses Hamulus pyramidis. Hamulus is an inconstant landmark. It is distant to the dural penetration point. The position of this point is given to Fig. 77.

Transnasal routes (Figs. 78 and 79)

A. carotis int. is the most important structure to identify in basal endoscopic approaches. Along the midline, approaches are less dangerous than lateral to it, except in the area of the carotid siphon. The anatomy of the siphon is well known from pituitary surgery. The next difficult segment is the carotid course to Foramen lacerum. This part of A. carotis int. is located more medially than other segments of the carotid artery. The endoscopic approach from Apertura nasi to the contralateral Apertura ext. of the carotid canal, is shown in Figs. 78 and 79. Foramen lacerum is located close to a straight lined direction. Foramen lacerum is the only area at the skull base where the artery crosses without a bony envelope, as it is usually given by the sphenoid bone and by the carotid channel of Pyramis. Therefore Canalis pterygoideus (Vidii) is used in modern endoscopy as an important landmark (Kassam, Snyderman et al. 2005). The canal connects the posterior basal segment of the wall of Sinus sphenoidalis to the area of Foramen lacerum and Apertura int. of the carotid canal. The nerve of Canalis pterygoideus (Vidii) – N.petrosus – should be spared.

SPECIAL SURGICAL ASPECTS. EXAMPLES.
(Figs. 71 to 79)

PLANNING STRATEGIES. IMAGING TECHNIQUES. EXAMPLES.
(Figs. 71 to 74)

Fig. 71

Imaging techniques. Topographical overview

A Craniospinal area
B Interforaminal cadaver skull dissection according to Fig. 70

guiding structures
Canalis nn.XII

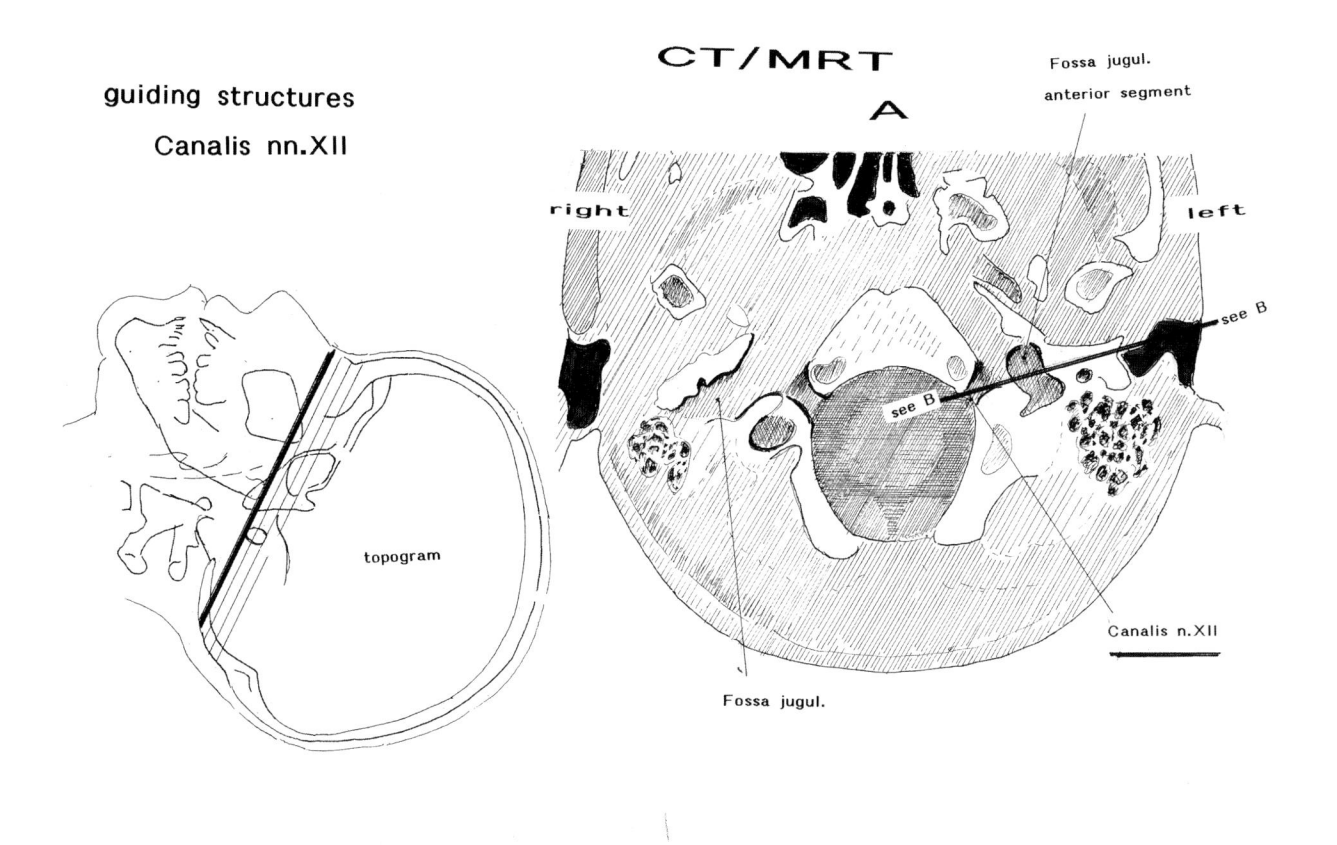

CT/MRT

A

right   left

Fossa jugul.
anterior segment

see B

see B

Canalis n.XII

Fossa jugul.

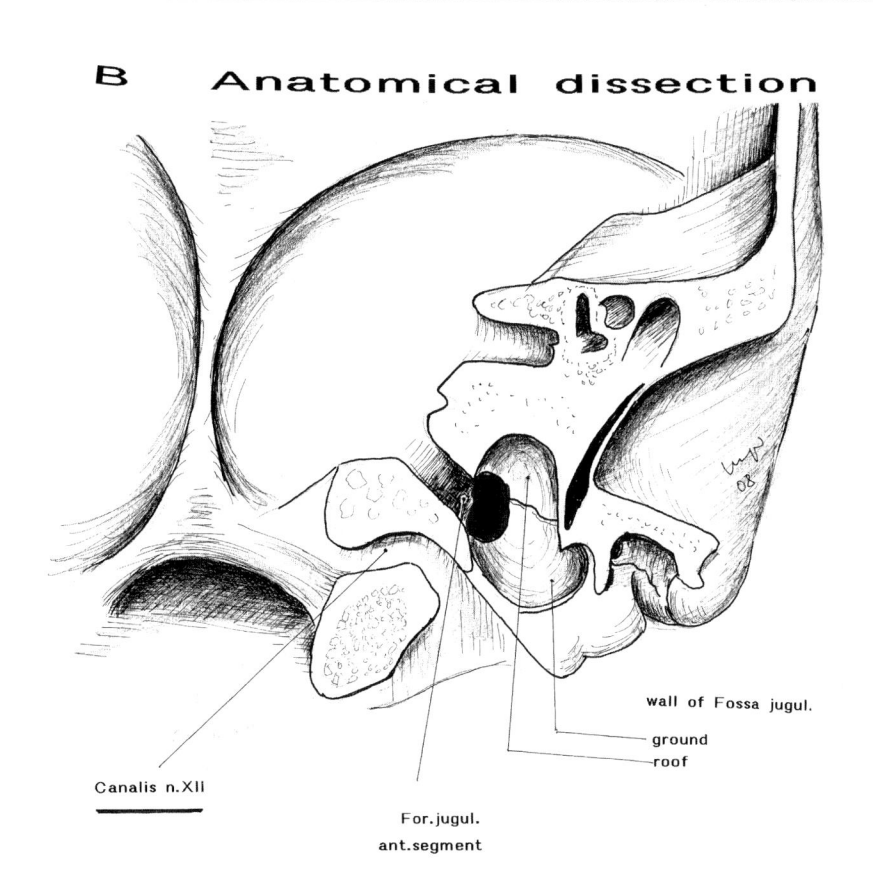

B   Anatomical dissection

Canalis n.XII

For. jugul.
ant.segment

wall of Fossa jugul.
ground
roof

Fig. 72

Continuation of Fig. 71
Right Canalis nervi hypoglossi present, as in Fig. 71
Left canal not present

A	CT/MRT, transectional level B indicated in
A' and A"	Sectional enlargements of A
B	Cadaver skull transection

Abbreviations

1	Apertura interna canalis n. hypoglossi
2	Canalis n. hypoglossi
(2)	projection
3	Foramen occipitale
4	Fossa jugularis
5	Foramen jugulare, anterior segment
6	as 4, and Foramen jugulare, posterior segment
7	Canalis caroticus
(7)	Apertura ext. canalis carotici, projection indicates
8	Meatus acusticus ext.
9	Cavum tympani
10	Tuberculum jugulare
11	Condylus occipitalis
12	Processus styloideus and Foramen stylomastoideum
13	Capitulum mandibulae
14	Meatus acusticus internus
15	Cellulae mastoideae

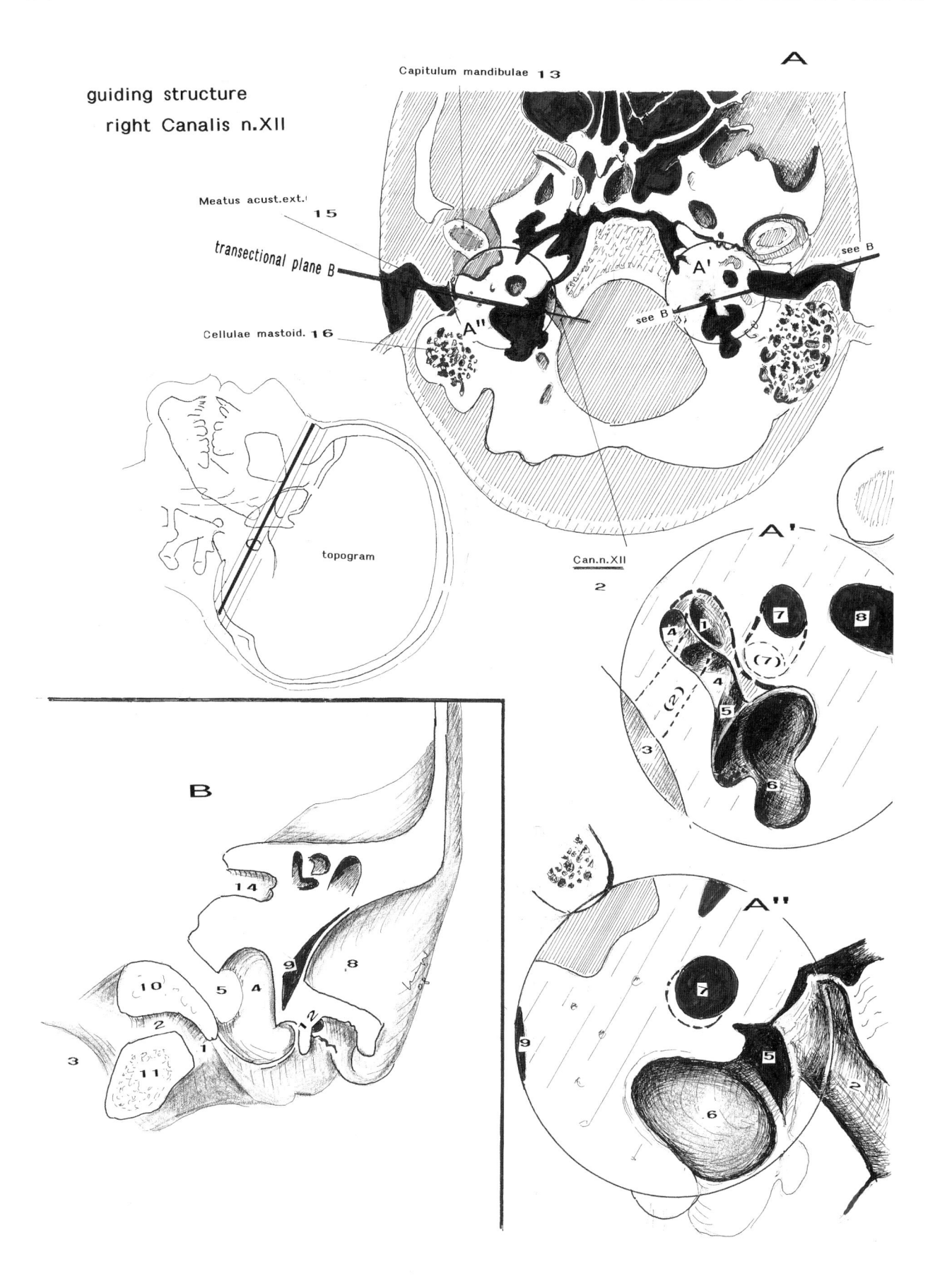

A

Capitulum mandibulae 13

guiding structure
right Canalis n.XII

Meatus acust.ext. 15

transectional plane B

Cellulae mastoid. 16

A'

A"

see B

see B

topogram

Can.n.XII
2

A'

B

14

Fig. 73

Continuation of Fig. 72
Different illustration of Capitulum mandibulae,
Cochlea, Foramen jugulare, Processus mastoideus, on the right and left side, in Figs. 72
and 73

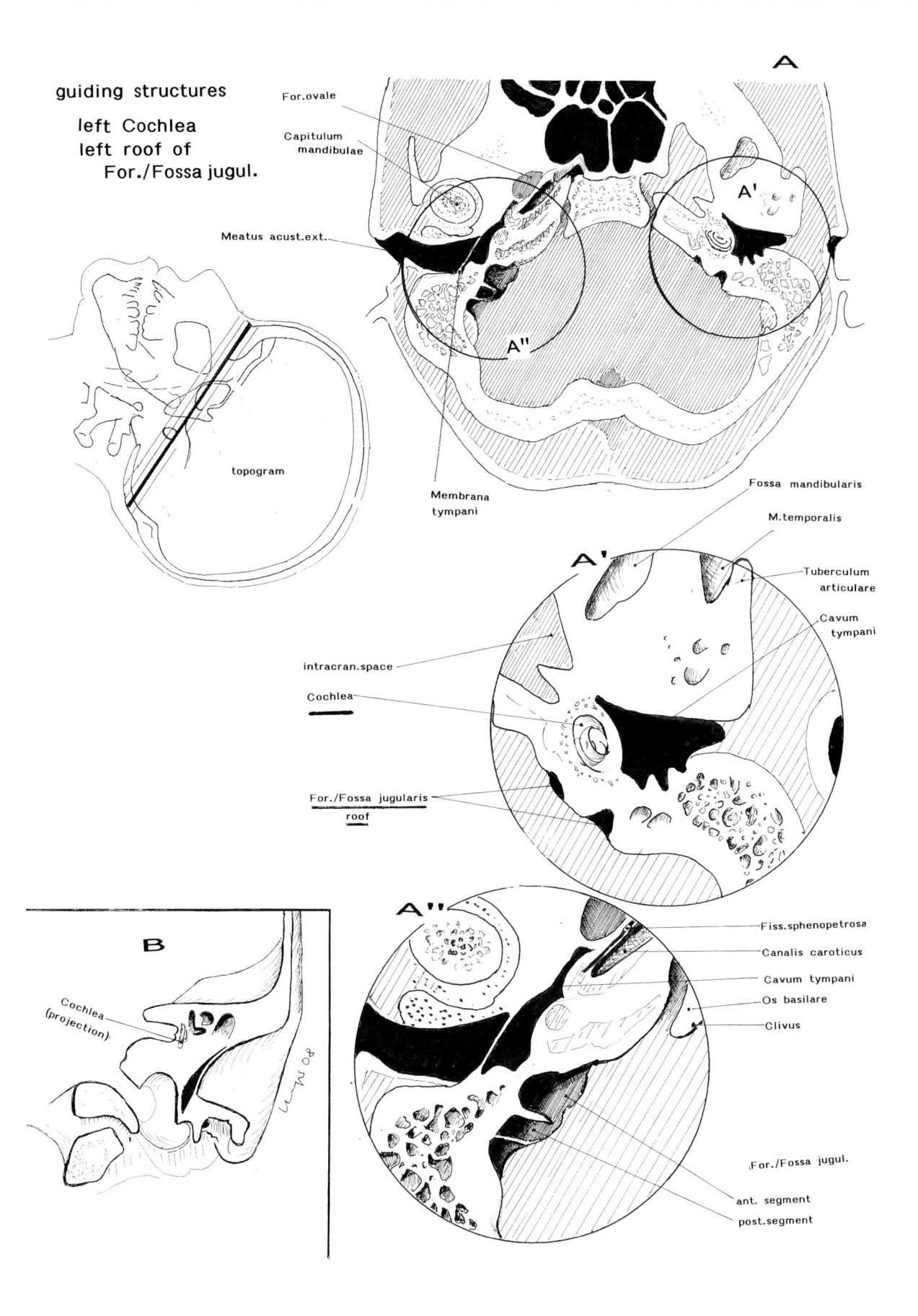

guiding structures

left Cochlea
left roof of
For./Fossa jugul.

For.ovale

Capitulum
mandibulae

Meatus acust.ext.

topogram

Membrana
tympani

Fossa mandibularis

M.temporalis

Tuberculum
articulare

Cavum
tympani

intracran.space

Cochlea

For./Fossa jugularis
roof

A

A'

A''

B

Cochlea
(projection)

Fiss.sphenopetrosa

Canalis caroticus

Cavum tympani

Os basilare

Clivus

For./Fossa jugul.

ant. segment

post.segment

Fig. 74

Continuation of Fig. 73
Different illustrations of Cochlea, Ducti semicirculares, and Meatus acusticus internus
on the right and left side, in Figs. 73 and 74.

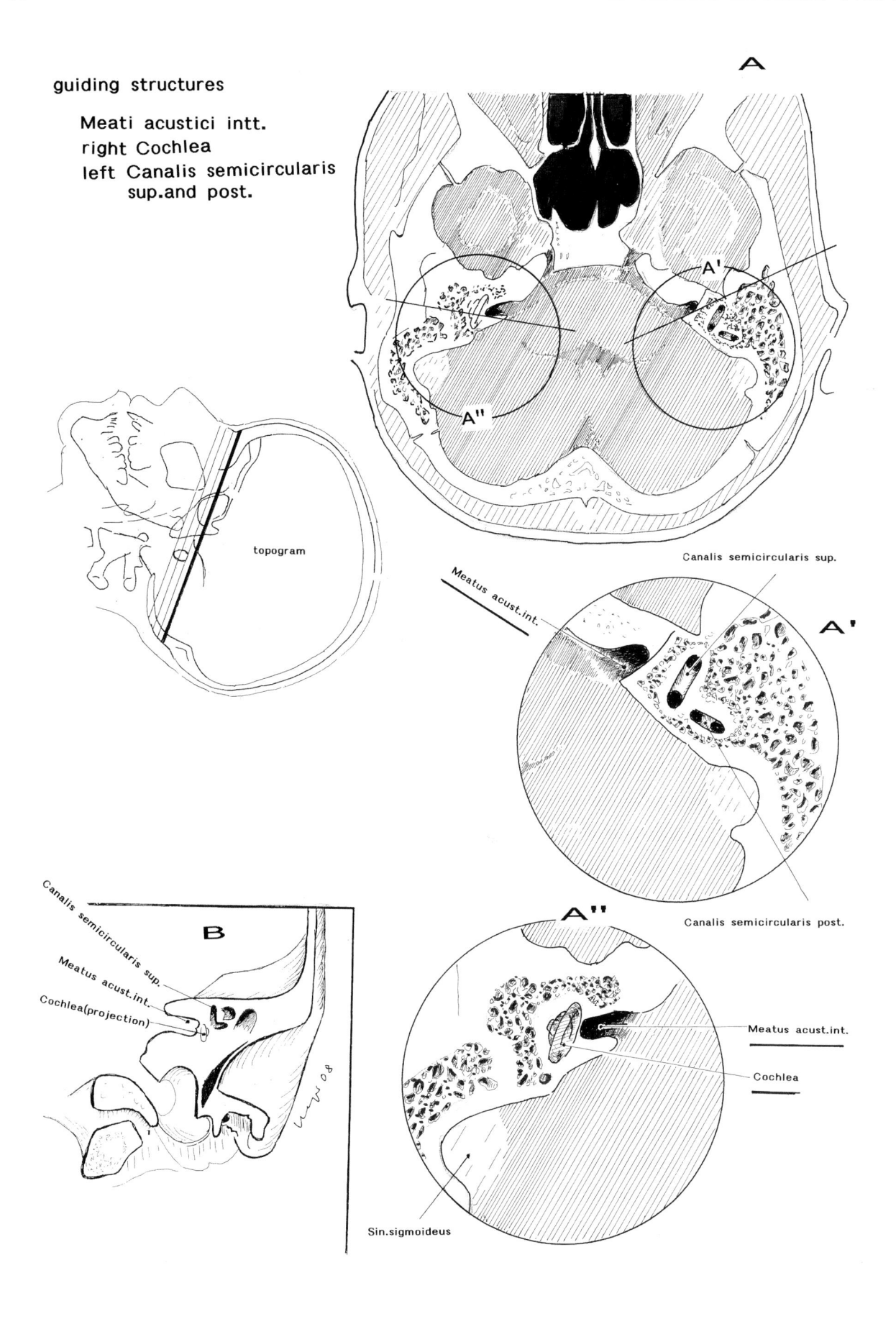

guiding structures

Meati acustici intt.
right Cochlea
left Canalis semicircularis
sup. and post.

topogram

A

Canalis semicircularis sup.

Meatus acust. int.

A'

Canalis semicircularis post.

B

Canalis semicircularis sup.
Meatus acust. int.
Cochlea(projection)

A''

Meatus acust. int.

Cochlea

Sin. sigmoideus

MEASUREMENTS (Figs. 75 to 77)

Fig. 75

Clivus area, underlying structures included.
Condylar-Tuberculum jugulare-complex.
Measurements based on 8 cadaver skull-dissection.

A and **A'**   Intracranial shape
B and **B'**   Extracranial shape

a   16–18 mm
b   20–25 mm
c   45 mm
d   50–55 mm
e   15–20 mm
f   25–35 mm
poc Synchondrosis petrooccipitalis

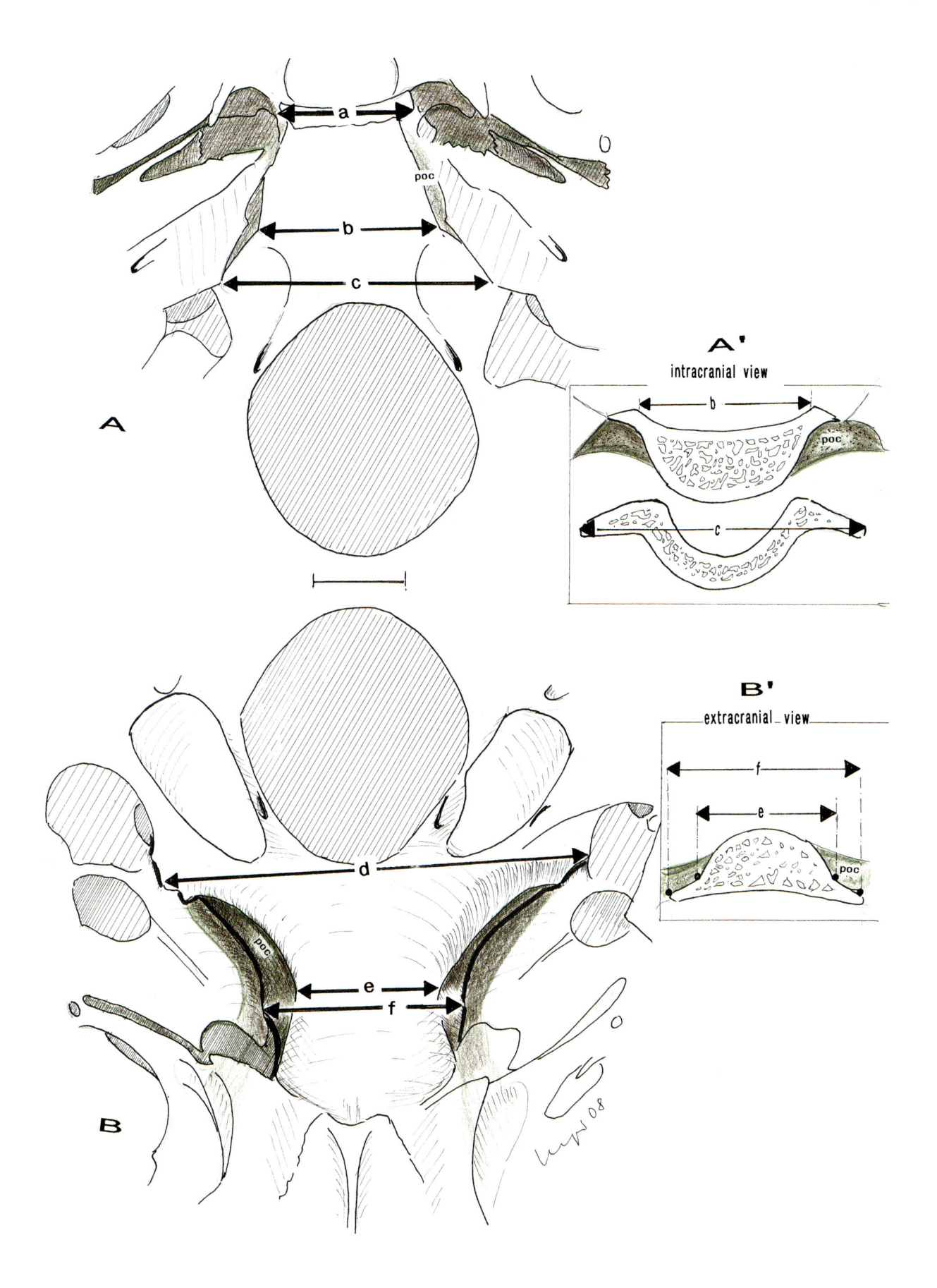

Fig. 76

Addendum to Fig. 75

Short distance of Canalis caroticus (A. carotis int.) to Fissura sphenopetrosa (bed of Tuba auditiva) and to Foramen ovale (N. mandibularis)
Enlargement of Sinus petrosus inf.

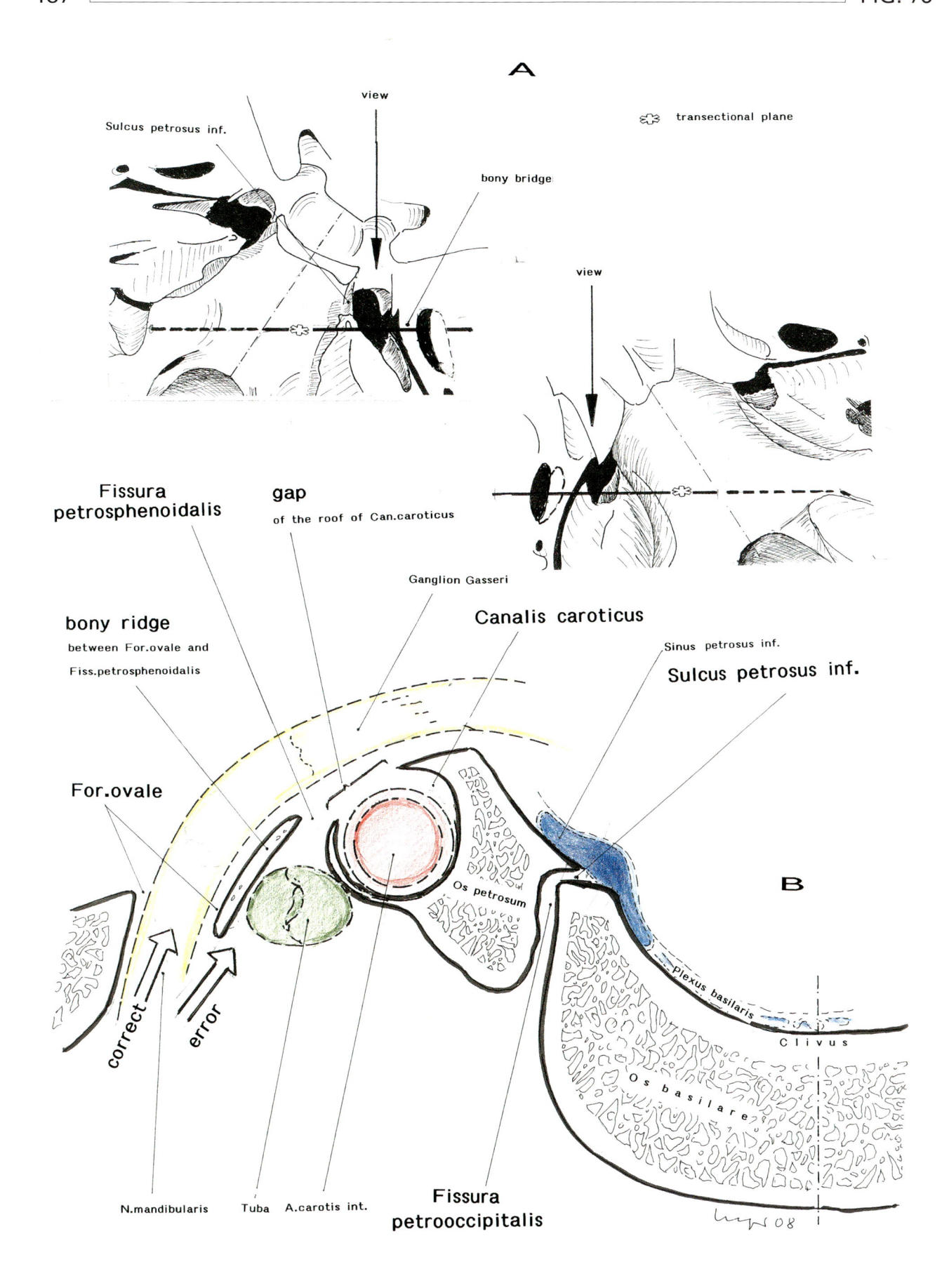

A

view

Sulcus petrosus inf.

bony bridge

transectional plane

view

Fissura
petrosphenoidalis

gap
of the roof of Can.caroticus

Ganglion Gasseri

Canalis caroticus

bony ridge
between For.ovale and
Fiss.petrosphenoidalis

Sinus petrosus inf.

Sulcus petrosus inf.

For.ovale

Os petrosum

B

Plexus basilaris

correct

error

Clivus

Os basilare

N.mandibularis   Tuba   A.carotis int.

Fissura
petrooccipitalis

hp 08

Fig. 77

Dural penetration point of N. abducens

The penetration point is often located at the medial margin of Sulcus petrosus inf. The sulcus is often flattened or variable. The length of Clivus varies between 3 and 4 cm but the relationship of segments a and b are 1 : 1,5. The distance to the midline is 7 – 9 mm. These measurements are useful for defining neuronavigatory landmarks.

FIG. 77

usual findings

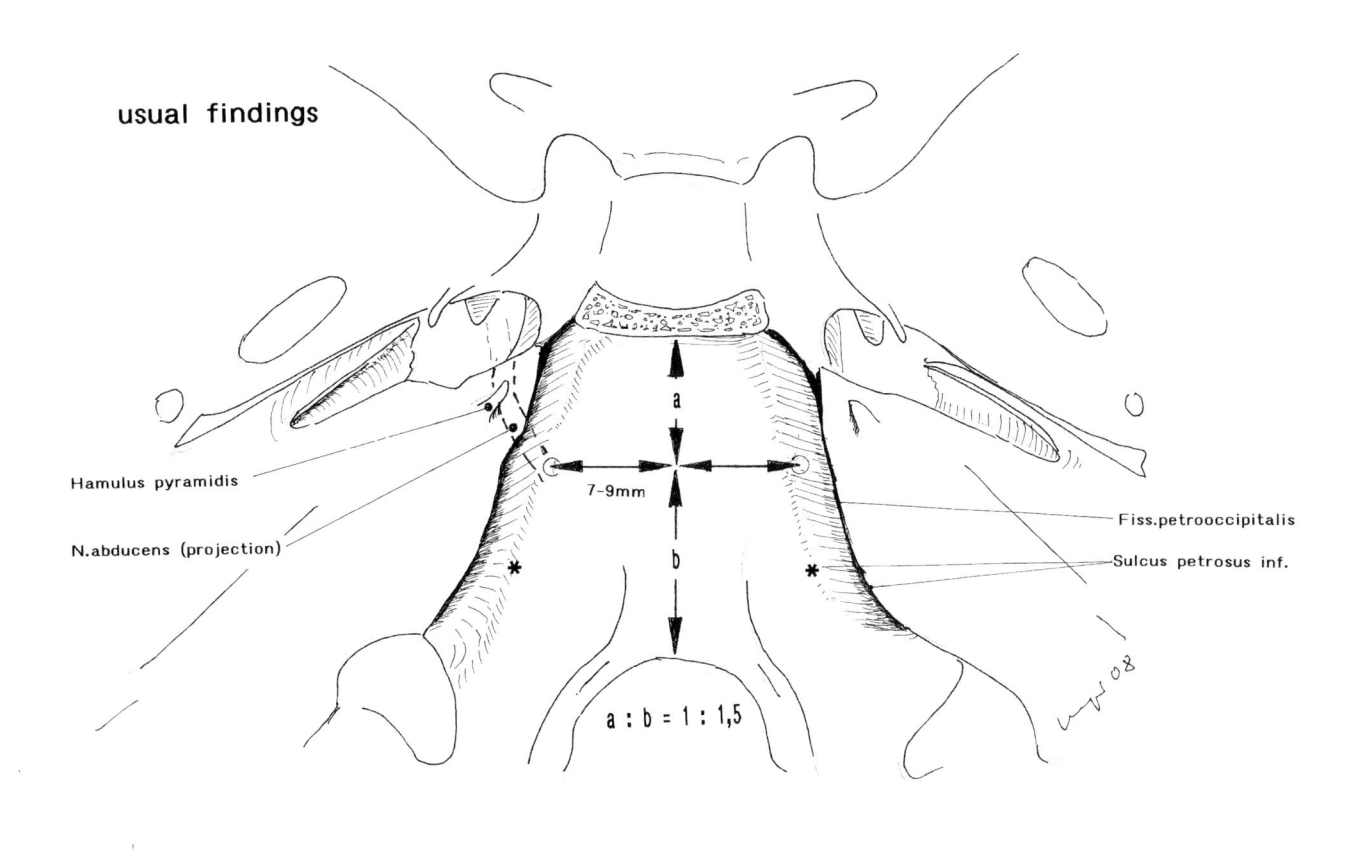

Hamulus pyramidis

N.abducens (projection)

7-9mm

Fiss.petrooccipitalis

Sulcus petrosus inf.

a : b = 1 : 1,5

✱ medial border area of Sulcus (not Sinus) petrosus inf.

◯ dural exit area of N. abducens

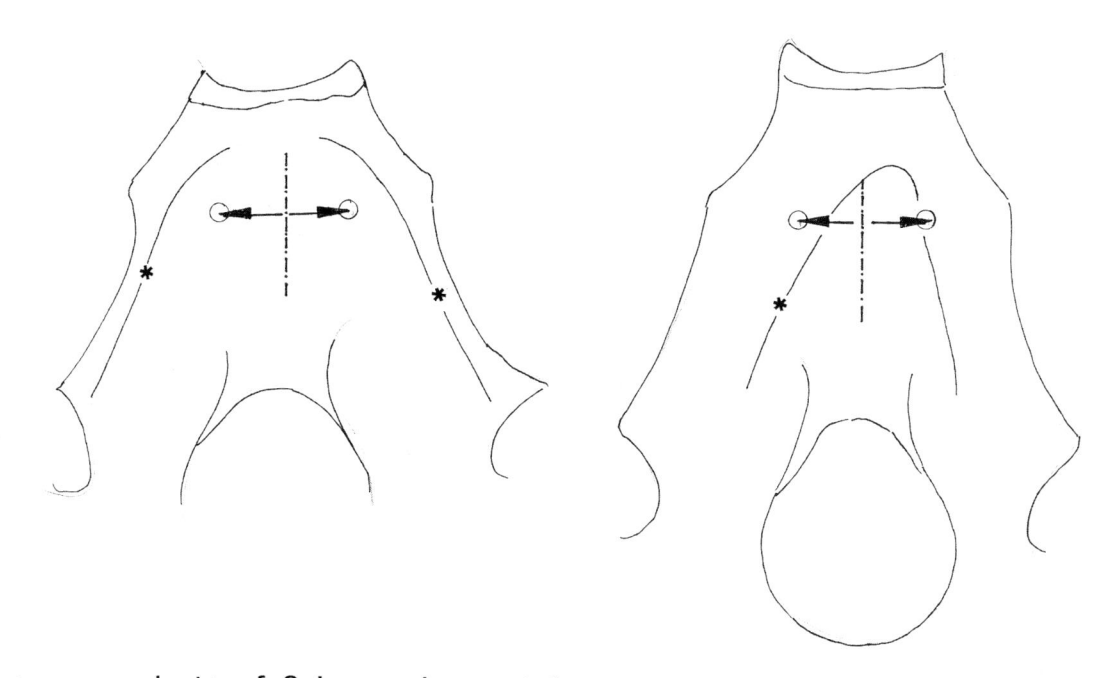

extreme variants of Sulcus petrosus inferior

Fig. 78

Endoscopic transnasal route
Anatomical model

Approach to the contralateral medial margin of Apertura externa canalis carotici. The medial margin of Foramen lacerum (according to the medial margin of the carotid artery) is crossed by the endoscopic route. Asymmetries of the skull base must be taken into consideration.

a and b: distances from to the midline for the definition of asymmetries. Asymmetric variant in B3 of Fig. 79

Abbreviations
1   Spina angularis (and Ostium tubae)
2   Foramen spinosum
3   Foramen ovale

FIG. 78

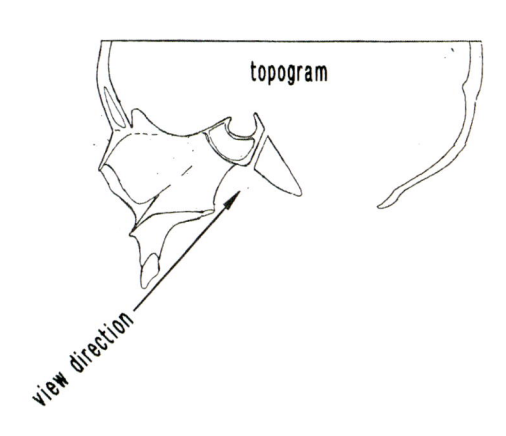

topogram

view direction

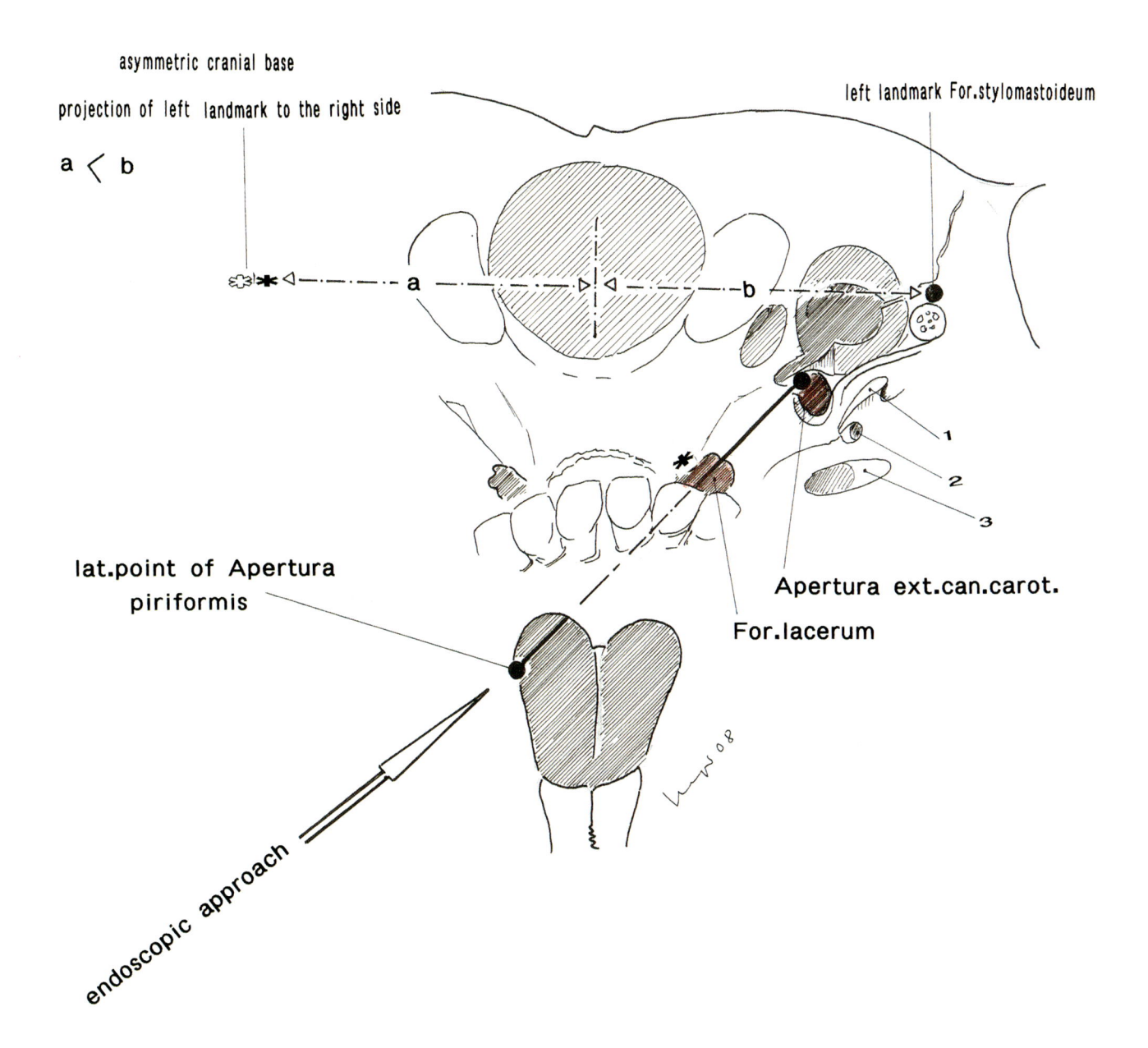

asymmetric cranial base

projection of left landmark to the right side

left landmark For.stylomastoideum

a ⟨ b

a

b

lat.point of Apertura
piriformis

Apertura ext.can.carot.

For.lacerum

1
2
3

endoscopic approach

Fig. 79

Addendum for Fig. 78

A   As Fig. 78
B to **B4**   further cadaver skull dissection

FIG. 79

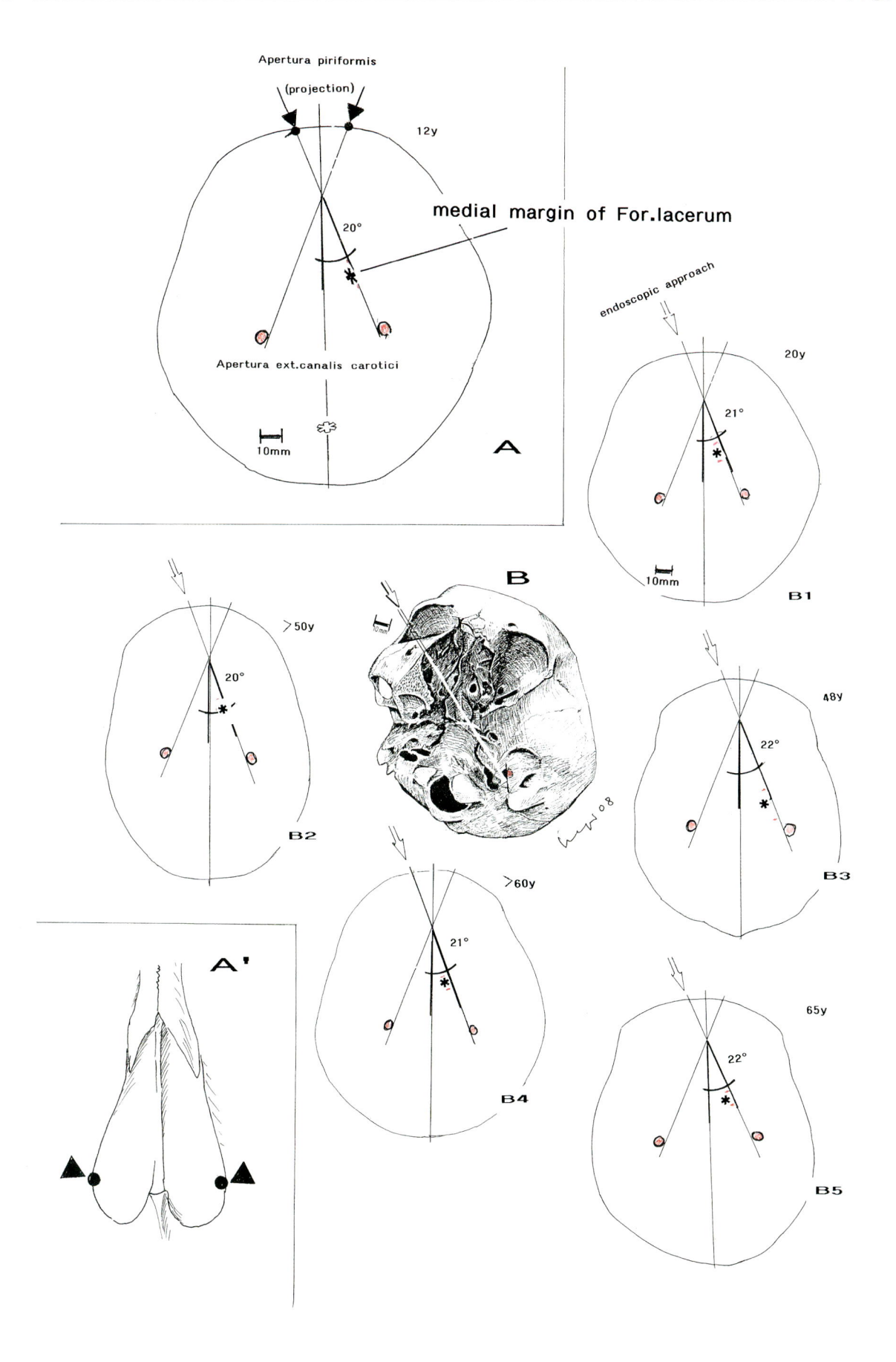

Apertura piriformis

(projection)

12y

medial margin of For.lacerum

20°

Apertura ext.canalis carotici

10mm

A

endoscopic approach

20y

21°

10mm

B1

>50y

20°

B2

B

48y

22°

B3

A'

>60y

21°

B4

65y

22°

B5

Literature

Cavallo LM, Messina A, Cappabianca P, Esposito F, de Divitis E, Gardner P, TSchabitscher M (2005). Endoscopic endonasal surgery of the midline skull base. Anatomical study and clinical considerations. Neurosurg Focus 19 (1), E2, pp 1–14

Corlieu P, Aaron C, Godefrov D (1981) Radioanatomie de 1a base du crâne, étage moyen et postérius. Ann Otolaryngol Chir Cervicofac 98, 173–179

Kassam A, Snyderman CH, Mintz A, Gardner P, Carrau RL (2005) Expanded endonasal approach: The rostrocaudal axis. Part II. Posterior clinoides to the foramen magnum. Neurosurg Focus Jul 15; 19(1): E4

Kassam AB, Gardner P, Snyderman C, Mintz A, Carrau R (2005) Expanded endonasal approach: Fully endoscopic, completely transnasal approach to the middle third of the clivus, petrous bone, middle cranial fossa and infratemporal fossa. Neurosurg Focus Jul; 15:19 (1):E6

Lang J (1979) Kopf, Teil B, Gehirn- und Augenschädel. Springer, Berlin Heidelberg New York

Lang J (1981) Neuroanatomie der Nn. opticus, trigeminus, facialis, glossopharyngeus, vagus, accessorius und hypoglossus. Arch Otorhinolaryngol 231, 1–69

Lang J, Schafhauser O, Hoffmann S (1983) Über die postnatale Entwicklung der transbasalen Schädelpforten: Canalis caroticus, Foramen jugulare, Canalis hypoglossalis, Canalis condylaris und Foramen magnum. Anat Anz 153, 315–357

Samii M, Janette PJ (1981) The cranial Nerves. Springer, Berlin Heidelberg New York

Subject Index

SpringerMedicine

Marc Sindou (Ed.)

Practical Handbook of Neurosurgery

From Leading Neurosurgeons

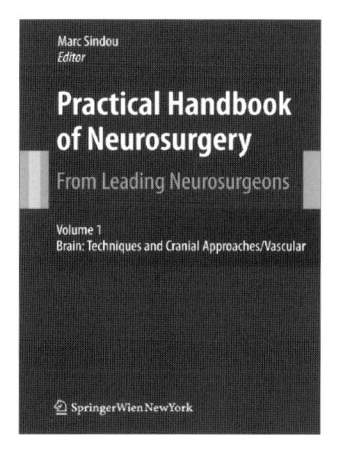

2009. XVIII, 1.695 p. 600 illus.
In 3 volumes, not available separately.
Hardcover. **EUR 499,–**
ISBN 978-3-211-84819-7

The book invites the reader to an exciting journey through the vast fields of neurosurgery, accompanied by a large panel of leading neurosurgeons. At a time when neurosurgery has a tendency to segment in many subspecialties, the goal was to regroup practical lessons from experienced neurosurgeons.

In addition, the book represents an anthology of ninety worldwide recognized neurosurgeons, with the main features of their curriculum and contributions.

The book has three volumes which cover the following items:

Volume 1: Techniques and cranial approaches; Vascular lesions; Cranial traumas; CSF/infectious diseases
Volume 2: Intracranial tumors; Intraoperative explorations; Pediatrics
Volume 3: Spine; Functional neurosurgery; Peripheral nerves; Education

The authors deliver their critical views and give useful guidelines.

SpringerWienNewYork

P.O. Box 89, Sachsenplatz 4–6, 1201 Vienna, Austria, Fax +43.1.330 24 26, books@springer.at, springer.at
Haberstraße 7, 69126 Heidelberg, Germany, Fax +49.6221.345-4229, SDC-bookorder@springer-sbm.com, springer.com
P.O. Box 2485, Secaucus, NJ 07096-2485, USA, Fax +1.201.348-4505, service@springer-ny.com, springer.com
All errors and omissions excepted. Recommended retail price. Net-price subject to local VAT.

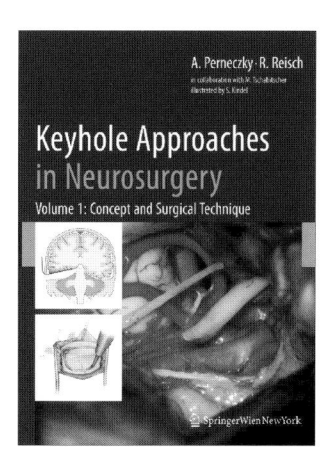

Th. P. Naidich et al.

Duvernoy's Atlas of the Human Brain Stem and Cerebellum

High-Field MRI, Surface Anatomy, Internal Structure, Vascularization and 3D Sectional Anatomy

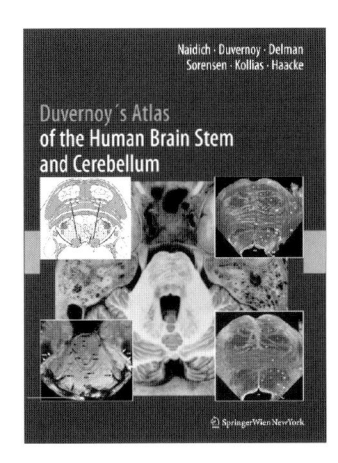

2009. XII, 876 p. With numerous illus.
Hardcover. **EUR 299,–**
ISBN 978-3-211-73970-9

Advanced MRI requires advanced knowledge of anatomy. This volume correlates thin-section brain anatomy with corresponding clinical 3 T MR images in axial, coronal and sagittal planes to demonstrate the anatomic bases for advanced MR imaging. It specifically correlates advanced neuromelanin imaging, susceptibility-weighted imaging, and diffusion tensor tractography with clinical 3 and 4 T MRI to illustrate the precise nuclear and fiber tract anatomy imaged by these techniques. Each region of the brain stem is then analyzed with 9.4 T MRI to show the anatomy of the medulla, pons, midbrain, and portions of the diencephalonin with an in-plane resolution comparable to myelin- and Nissl-stained light microscopy (40-60 microns).

The volume is carefully organized as a teaching text, using concise drawings and beautiful anatomic/MRI images to present the information in sequentially finer detail, so the reader easily assimilates the relationships among the structures shown by high-field MRI.

SpringerWien NewYork

P.O. Box 89, Sachsenplatz 4–6, 1201 Vienna, Austria, Fax +43.1.330 24 26, books@springer.at, springer.at
Haberstraße 7, 69126 Heidelberg, Germany, Fax +49.6221.345-4229, SDC-bookorder@springer-sbm.com, springer.com
P.O. Box 2485, Secaucus, NJ 07096-2485, USA, Fax +1.201.348-4505, service@springer-ny.com, springer.com
All errors and omissions excepted. Recommended retail price. Net-price subject to local VAT.